Roofing Fundamentals

An Overview of the Roofing Industry

Developed with funding provided by
The Roofing Alliance

© 2025 Nieri Department of Construction and Real Estate Development
and the Roofing Alliance

ISBN 978-1-63804-182-5
eISBN 978-1-63804-183-2

Distributed by Clemson University Press

Roofing Fundamentals

An Overview of the Roofing Industry

Dr. Dhaval Gajjar

Dr. Jason Lucas

Editors

Dhaval Gajjar, Ph.D., FMP, SFP

Dr. Dhaval Gajjar is an Assistant Professor at Clemson University's Department of Construction Science and Management in the College of Architecture, Arts and Construction. Dr. Gajjar has significant construction industry experience working as a Project Manager for both the general contractor and an owner organization responsible for large remodel, renovation, and TI projects. Dr. Gajjar also has 10+ years of significant research experience related to workforce development, education and training, project delivery and performance measurement. He has authored over twenty (20+) refereed journal publications, one book chapter, and over fifteen (15+) conference presentations disseminating the research results. He has also conducted over fifty (50) industry presentations educating industry professionals on using the research tools. He is also a recipient of Bennett Award and is a certified Facility Management Professional (FMP) and Sustainable Facility Professional (SFP).

Jason Lucas, Ph.D.

Dr. Jason Lucas is an Associate Professor in the Department of Construction, Development and Planning at Clemson University. He holds a Bachelor of Architecture degree from the New Jersey Institute of Technology and a master's in Building Construction Science and Management and a PhD in Environmental Design and Planning from Virginia Tech. Dr. Lucas has conducted research in workforce development, online education, and the use of technology in education. He has over 10 years of research and teaching experience at Clemson University and has published over twenty (30+) journal articles and over thirty (30+) conference presentations.

Acknowledgements

Assistant Editor:

Gopika Viswanathan

Gopika Viswanathan is a graduate student in the Masters of Construction Science and Management program. She also works as a Research and Teaching Assistant at Clemson University.

Layout Designers:

Danett Vargas Sanchez

Danett Vargas Sanchez is an undergraduate student double majoring in Art and Communication at Clemson University. She is part of the Clemson University Honors College and has four years of graphic design and fine art experience.

Amelia Lyles

Amelia Lyles is an undergraduate student majoring in Graphic Communications at Clemson University. She also works as a Teacher Assistant for Clemson and has two years of graphic design experience.

Technical Reviewers:

William Good, Roofing Alliance
Mark Graham, NRCA

Roofing Alliance Staff:

Alison L. LaValley, CAE
Jessica Priske, Director
Nicole Christodoulou, Manager
Maggie Kosinski, Manager

Acknowledgements

Contributors:

Bill Good, Roofing Alliance

Dudley Miles, J.D. Miles & Sons, Inc.

Bruce McCrory, Nations Roof

David Landis, Petersen Aluminum Corporation

Kyle Cahill, King of Texas Roofing

David Allor, OMG Roofing Products

John Thomas, Siplast, Inc.

Andrew Christ, Cornell Roofing & Sheet Metal Company

Dan Jarratt, Piper Roofing

Justin Reeder, REI Engineers

Jeremiah Price, Soprema

Candace Klein, Klein Contracting Corporation

Robert Pringle, CHST, CSE, CECM, NYSCSC, Evans Roofing Company, Inc

Chris Huettig, KARNAK Corporation

Piers Dormeyer, EagleView

Preface

This series of manuals on the roofing industry is made possible through the support of the Roofing Alliance and numerous industry professionals. The manuals were developed from lectures given by industry experts and are meant as an introduction to the topics. It is not intended to include the entirety of the industry and the information about all types of roofing products or systems. For a more complete discussion of roofing materials and installation methods, the reader is encourage to see the NRCA Roofing Manual, available from the National Roofing Contractors Association (www.nrca.net).

The information contained has been reviewed for technical accuracy and clarity at the date of its publication. Codes and practices change over time, so the editors intend to periodically review, revise, and publish future editions of these manuals to reflect those changes.

The roofing manual series can be used independently to provide an overview of different parts of the roofing industry but also serve as a complimentary summary of the knowledge presented in the Clemson Online Professional Development series of courses that have been created with the support of the Roofing Alliance:

• Roofing Fundamentals
• Roofing Management
• Roofing Business and Leadership

The content is broken up into three topic areas to allow tailored focus of employees and business leaders to focus on the area most relevant for them. Roofing Fundamentals provides a general overview of the industry, the products and services available, and focuses on developing an understanding of systems and terminology. Roofing Management covers topics related to a project, including codes, scheduling, field crew management, quality control, risk management, and site logistics. Finally, Roofing Business and Leadership delves into leadership strategies, sales, marketing, and various aspects of owning a business within the roofing industry.

This project would not have been possible without the numerous industry supporters who have donated time and content in support of creating these educational resources.

Table of Contents

Table of Contents

Table of Contents

Table of Contents

Table of Contents

CHAPTER

01

Introduction to Roofing

With content donated by
William Good
Roofing Alliance

1.1 Introduction

When most people think about roofs, the first thing that comes to mind is either the roof on their own home or a famous roof, such as the one on the Chrysler Building in New York City. The obvious reality is that every built structure has a roof, and the purpose of this Manual is to describe why they are important, how they should be designed and built, and how companies operate successfully in the roofing industry.

This Manual will also discuss different jobs in the roofing industry and career opportunities for those who choose to enter it. The roofing industry is estimated to employ more than 250,000 people, and the projections are that more people will be needed for the foreseeable future.

1.2 The Basics

Very broadly, roofs fall into two categories: steep-slope and low-slope, as shown in Figure 1.2.1. The figure also shows the common angles that are typically found for a roof. Steep-sloped roofs generally have a slope of 2:12 or greater, and are water-shedding. Low-sloped roofs have slopes of less than 2:12 and are seen mostly on commercial and industrial buildings. The industry does not refer to them as "flat" roofs because the roof should always have positive slope to provide for drainage.

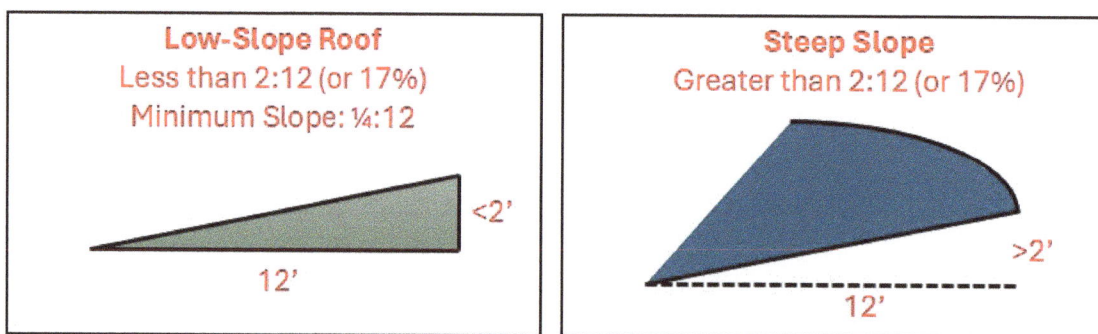

Low-Slope Roof	Steep Slope
Less than 2:12 (or 17%)	Greater than 2:12 (or 17%)
Minimum Slope: ¼:12	
<2'	>2'
12'	12'

Figure 1.2.1 – Roof Slopes
Source: Jason Lucas

1.3 Steep-Slope Roofing

Steep-slope roofs are generally defined as having a slope of 2:12 or greater. They are typically found on houses, small commercial buildings and churches.

The most common steep-slope roofing material is asphalt shingles, which compose about 80% of the steep-slope market. Other materials include metal, slate, copper, clay tile, concrete tile and wood shingles and shakes. An example of a standard asphalt shingle (left) and architectural shingles (right) have been displayed in Figure 1.3.1 below.

Figure 1.3.1 - Asphalt Shingles
Source: Editor

Steep-Slope Configurations

There are five basic configurations of steep-slope roof systems, as shown in Figures 1.3.2 through 1.3.6. They are a hip roof, gable roof, mansard roof, shed roof and gambrel roof.

- A hip roof slopes downward from the peak on all sides.
- A gable roof consists of two sections whose upper horizontal edges meet to form a ridge.
- A mansard roof has two slopes on each side; the lower slope is much steeper than the upper slope.
- A shed roof is a steep-slope roof on a steep plane.
- A gambrel roof is a symmetrical two-sided roof in which each side has two slopes: one steeper and one shallower.

Figure 1.3.2 – Hip Roof
Pexel: Pixabay (CC0)

Figure 1.3.3– Gable Roof
Pexel: Alexander Isreb (Creative Commons)

Figure 1.3.4 – Mansard Roof
Pexel: Maria Orlova (Creative Commons)

Figure 1.3.5 – Shed Roof
Pexel: Expect Best (Creative Commons)

Figure 1.3.6 – Gambrel Roof
Pexel: Azra (Creative Commons)

A ridge is the topmost part of the roof. A valley is the area where two or more planes of the roof intersect. Fascia is the vertical band under a roof edge, or the outer surface of a cornice that is visible to an observer. The soffit is the material beneath the eave that connects the far edge of the roof to the exterior wall of the house or building (See figure 1.3.7)

Figure 1.3.7 - Roof Terminology

1.4 Low-Slope Roofing

Low-slope roofs make up about 2/3 of the roofing market. Low-slope roofs tend to be more complicated that steep-slope roofs, because of a number of design and building code requirements.

In the last half of the 19th century, industrialization created the need for larger buildings, and low-slope roofs were found to be much more cost-effective than traditional steep-slope roofs. The origin of built-up roofing can be traced to Chicago, where the gas companies were giving away a waste product called coal tar pitch.

Coal tar pitch was a good waterproofing material, but it had viscous properties (it would melt when heated), so roofing felts – or "tar paper" were introduced as a way to provide stability to the coal tar pitch. That roof system – alternating layers of coal tar pitch and roofing felts – became known as built-up roofing.

Later, asphalt – a residue of the oil refining process – was found to have properties similar to coal tar pitch, and was used more commonly in built-up roofs. From the late 19th century to about 1980, built-up roofs dominated the low-slope roofing market.

Beginning in the 1970s, a number of factors led to a decline in the popularity of built-up roofs. For one thing, oil prices escalated dramatically, most notably after the oil embargo of of the mmid-1970s. Health concerns were also raised about asphalt and coal tar pitch fumes, and in fact coal tar pitch was classified as a known human carcinogen, or cancer-causing agent.

As a result, new products were introduced into the low-slope roofing market; these collectively became known as "single-ply" roofing products because they were applied in a single layer. These new materials included the following:

- Ethylene propylene diene monomer (EPDM), a product that looks and behaves like rubber
- Polyvinyl chloride (PVC), a product used commonly for such things as shower curtains
- Thermoplastic polyolefin (TPO), which looks and feels like PVC, but has a different chemical composition

At about the same time, a product called modified bitumen, originating in Western Europe, entered the market. This product uses polymer-modified asphalt with roofing felts, pre-manufactured in a factory and applied in a single layer or, often, in two layers.

At about the same time, a product called modified bitumen, originating in Western Europe, entered the market. This product uses polymer-modified asphalt with roofing felts, pre-manufactured in a factory and applied in a single layer or, often, in two layers.

Additionally, metal roof systems have become an important segment of the low-slope (and steep-slope) roofing market. And in some parts of the country, sprayed-in-place polyurethane foam roof systems perform very well.

Installing a Low-Slope Roof

Among the factors to be considered when choosing and installing a low-slope roof are the following:

- Building codes. The three primary topics covered by building codes are wind resistance, fire resistance and energy conservation. These will be discussed in more detail later in this Manual.

- Design. Because low-slope roofs are not water-shedding, drainage, attachment, flashings and penetrations must all be addressed as the roof is being designed.

- Safety. Falls are the biggest source of severe accidents and fatalities in the roofing industry, but roof safety also includes being aware of a host of other hazards that may be present on the roof.

- Installation. The roof must be properly attached to the substrate and have seams properly sealed, among other things.

- Logistics. Material handling, especially in urban areas, can be a challenge and must be carefully considered.

1.5 How Big is the Roofing Industry in the U.S.?

According to a study completed by researchers at Arizona State University, there are about 50,000 roofing contractors in the U.S. The same study concluded that the roofing industry could employ as many as 1 million people, including those working in all segments of the industry.

Most roofing contractors are privately owned; they range in size from having annual sales of about $1 million to having annual sales over $1 billion. The low-slope roofing material manufacturing sector has four companies that combine for about 60% of total sales in that sector: GAF, Holcim, Johns Manville and Carlisle SynTec. The asphalt roofing sector includes several companies: Owens-Corning, CertainTeed, GAF, IKO, Malarkey Roofing Products and TAMKO. Many manufacturers of roofing materials also produce roof insulation and accessories. Other companies provide such things as roof equipment and fasteners.

The wholesale distribution segment of the industry is vital because most roofing materials are sold through distribution rather than directly to contractors. The distribution segment has three large companies with national footprints: ABC Supply, Beacon Supply and SRS Distribution.

1.6 The Evolution of the Roofing Industry Continues

Prospects for the future of the roofing industry are very good, for several reasons. First, about 75% of the total roofing market – both steep-slope and low-slope – is composed of repair, maintenance, service and replacement. Just 25% is created by new construction. Second, there is a large amount of roof inventory that is about to reach the end of its expected life cycle. That inventory includes homes that were built 20-30 years ago, as well as a large number of warehouses and distribution centers that were constructed during the rise of online sales.

And roofs today are evolving to meet more demands than simply keeping water out of a home or building. For example:

- Roofs serve as platforms for mechanical and HVAC systems, which makes their installation and maintenance more important – and more complicated.

- Solar panels are increasingly being used on roofs as a source of energy; there are also some products in the market today that use embedded solar cells. That trend is expected to continue, although it will need to be managed carefully. For instance, solar panels increase the amount of heat that passes through to the roof system, which can cause it to age more quickly. Also, solar panels have a life expectancy of about 20 years and will need to be replaced. Roof systems, too, have a life expectancy of roughly 20 years, so owners need to be careful not to install solar panels on older roof systems (Fig 1.6.1).

- Because roofs can be the source of significant heat and energy loss, increasing amounts of roof insulation are included as part of low-slope roof assemblies.

- In some parts of the world, roofs are being reimagined as a means for capturing and reusing rainwater, for example, there are now "blue roofs" in Australia.

- Under the right conditions, roofs can also support vegetative growth, which not only adds aesthetic appeal but can also help to save energy, capture CO_2 emissions and mitigate stormwater runoff (Fig 1.6.2).

Figure 1.6.1 – Solar installation/integration
Pexel: Trinh Tran (Creative Commons)

Figure 1.6.2 – Vegetative Roof Systems
Photo by Unsplash from freerangestock.com

1.7 Careers in the Roofing Industry

A study done at Clemson University shows the variety of careers – and career paths – that exist in the roofing industry.

In the contractor segment, the most common career path for a graduate of a construction management program is to be an estimator, project manager or superintendent. All of those positions include some level of customer interaction as well as some part of project management.

Because many roofing companies are smaller than most general contractors, people entering the roofing industry will have the opportunity to advance much more quickly in those types of positions.

In the manufacturing segment, the most common entry position is in sales and marketing, selling to contractors, owners, developers and architects. Another common position is that of a technical representative; because manufacturers offer long-term warranties, it is important to have the roof inspected during its installation. Technical representatives also assist with any problems that may arise after installation.

In the distribution segment, the larger companies are highly decentralized and provide autonomy to branch managers. Distribution companies typically have training programs that lead to the position of branch manager, and later to regional or national management positions.

Other careers in the roofing industry include the growing profession of roof consulting. Roof consultants typically work for building owners and manage the entire roofing process for them; they also provide forensic work in the event of a roof problem or failure.

1.8 Who Decides What Roof System to Choose?

Because the large majority of roofing work does not involve new construction, the roofing contractor is often the prime contractor and deals directly with the home or building owner.

However, roof consultants also market their services to building owners, and often will recommend a particular roof system and/or a list of qualified contractors from which to solicit proposals. And roofing material manufacturers also try to influence the buying decision by working with architects, consultants and building owners.

The roofing industry is also unique in that most roof systems offer long-term warranties. Manufacturers who offer long-term warranties create programs for "licensed applicators" or "approved" contractors who can install warranted roofs.

So, the buying decision for a new roof can become complicated and suggests that personal relationships at every level are critically important.

CHAPTER

02

Steep-Slope Roof Systems

With content donated by
Dudley Miles
J.D. Miles & Sons, Inc.

13

2.1 Introduction

Chapter One covered the five basic types of structures that incorporate steep slope roof systems. In this chapter, we will go into much more detail about how steep slope roofs are designed and installed.

2.2 Components of a Steep-Slope Roof Assembly

The first component of a steep-slope roof assembly is the roof deck that is composed of plywood or OSB (Figure 2.2.1). The deck forms the bottom part of the roof system and is the platform from which the other components will be supported. The deck can be made of oriented strand board (OSB) (Figure 2.2.2) or wood plank (Figure 2.2.3). Battens may also be installed on top of the deck if needed. Fire requirements need to be considered when choosing a deck material. When required by building code, a fire-treated deck must be installed to avoid potential fire spread from one occupancy space to another, such as in adjoining townhomes.

Note that the roofing contractor is usually not involved with the installation of the roof deck, but must still be aware of code and other requirements.

Figure 2.2.1 – Plywood or OSB Sheathing Roof Deck
Source: J.D. Miles & Sons, Inc.

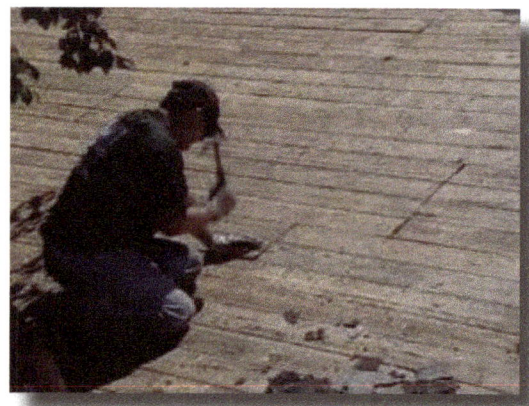

Figure 2.2.2– Wood Plank Deck
Source: J.D. Miles & Sons, Inc.

Figure 2.2.3 – Oriented Strand Board
(OSB) Deck
Source: J.D. Miles & Sons, Inc.

The other components of a roof assembly include insulation and an underlayment (Figure 2.2.4). Note that insulation is usually installed in the attic space formed by the structural framing under the roof deck. Underlayment can be made of felt paper or other synthetic materials that are installed on top of the roof deck. They help prevent water penetration and serve as a moisture barrier.

Figure 2.2.4 – Felt Paper as Underlayment
Source: J.D. Miles & Sons, Inc.

2.3 Insulation

Every steep-slope roofing system needs to have some type of insulation according to the standards and requirements of local building codes. There are two different types of roof insulation used in steep-slope roof systems: loose-fill insulation and batt insulation.

Loose-fill insulation consists of small particles of fiber, foam or other materials (Figure 2.3.1), while batt insulation is made of fiberglass or mineral wool that is precut into pieces (Figure 2.3.2). The insulation can also be installed above-deck that consists of rigid board insulation in a nail base configuration.

Figure 2.3.1 – Loose-fill Insulation
Source: J.D. Miles & Sons, Inc.

Figure 2.3.2 – Batt insulation
Source: J.D. Miles & Sons, Inc.

It is important to note that if the contractor is working on an existing house and the attic space has no insulation, the contractor should let the owner know. It is important to note that the code does not require attic insulation upgrades in typical reroofing situations. The amount of insulation, or total R-value, that is required, varies on the climatic zone where the structure is located.

2.4 Ice Dam Protection

Something to consider with steep-slope roofs is ice accumulation. In fact, about 70% of the U.S. has ice accumulation at some point during the year.

An ice dam protection membrane is important in the event of ice on a roof. The ice dam protection membrane is installed on top of the underlayment (Figure 2.4.1). After installing the ice dam protection membrane on the roof, a water-shedding roof covering (such as asphalt shingles) will be added to finish the assembly (Figure 2.4.2). Shingles are installed by overlapping units commonly known as the head lap. The head lap is used to avoid water penetrating inside the shingle. With the head lap, water will not have enough space to travel underneath the roof system.

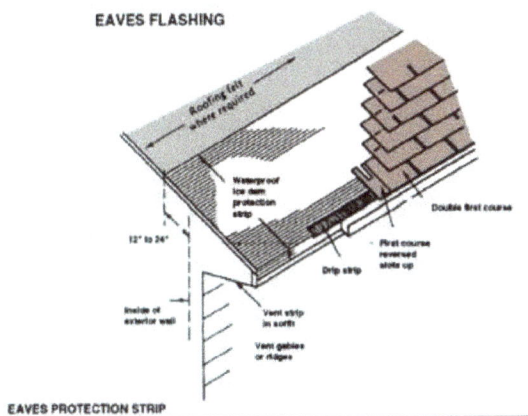

Figure 2.4.1– Installation of the ice dam protection membrane
Source: J.D. Miles & Sons, Inc.

Figure 2.4.2 – Installation of Asphalt Shingles
Source: Jason Lucas

A traditional shingle has three tabs (Figure 2.4.3). A newer product has also emerged in the roofing industry called laminated shingles (Figure 2.4.4). They are heavier than traditional "3-tab" shingles; the extra weight provides more protection from nature. Laminated shingles often give an appearance more like slate or wood. The technology for laminated shingles, as the name suggests, involves laminating two pieces of asphalt shingles together.

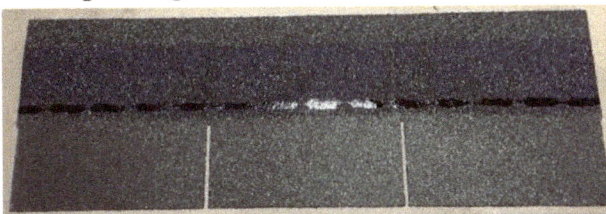

Figure 2.4.5 – Typical shingle
Source: J.D. Miles & Sons, Inc.

Figure 2.4.7 – Laminated shingle
Source: J.D. Miles & Sons, Inc.

2.5 Ventilation

Poor ventilation within the roof system can cause ice damming in colder climates or excessive condensation in hot and/or humid climates. To control the introduction of moisture into the roof system, ventilation, insulation and air barriers can all be used. Air barriers that are usually polyethylene sheet material are installed above the ceiling but below the rafters. All openings through the ceiling should be sealed to minimize air infiltration into the attic. Sprayed-in-place closed-cell foam can also be an effective air barrier. Insulation helps to keep interior space close to their ambient temperatures. When interior heat loss is minimized, the opportunity for condensation to occur is also minimized. Nails extending through the roof deck into the attic can cause condensation due to hot, moist air rising and meeting the colder metal temperatures on the outside. This can cause water droplets to form, leading to moisture damage.

Continuous ridge vents (Figure 2.5.1 & 2.5.2) combined with soffit ventilation allow outside air to flow into the entire attic and displace super-heated and moist air. Air is always able to exit the attic and exhaust contaminants from the home. Additionally, ridge vents are installed along the entire length of a ridge. Ridge vents are either "shingle over" or "non-shingle over" types.

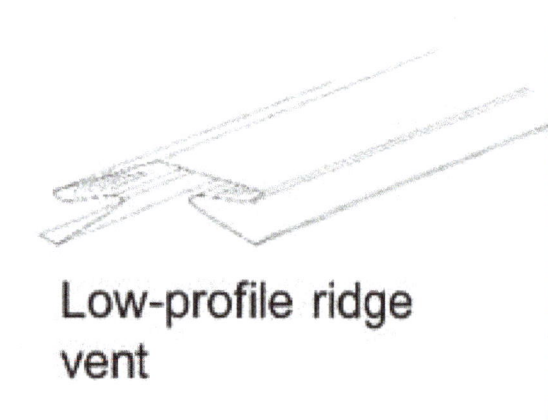

Figure 2.5.1 – Low Profile Ridge Vent
Source: J.D. Miles & Sons, Inc.

Figure 2.5.2 – Example of a Ridge Vent
Source: J.D. Miles & Sons, Inc.

Roof power vents, as shown in Figure 2.5.3, are also commonly used. A common problem found in steep-slope roofing is when a ridge vent is installed during a reroofing project while original static vents remain. This creates a "short circuit" to the convection process and nullifies air movement in low areas of the attic space. This must be avoided. A diagrammatic representation is shown in Figure 2.5.4 below.

Figure 2.5.3 – Roof Power Vent with Water Shedding Covering
Source: J.D. Miles & Sons, Inc.

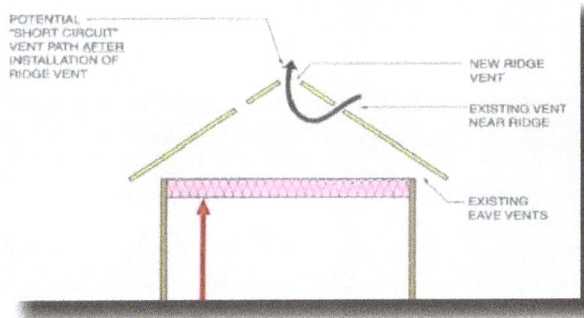

POTENTIAL "SHORT CIRCUIT" VENT PATH AFTER INSTALLATION OF RIDGE VENT

NEW RIDGE VENT

EXISTING VENT NEAR RIDGE

EXISTING EAVE VENTS

Figure 2.5.4 – Power Vent Issue
Source: J.D. Miles & Sons, Inc.

Other roof exhaust vents can be installed in the field of the roof or in the roof end gable (Figure 2.5.5).

Figure 2.5.5 – Exhaust Vents
Source: J.D. Miles & Sons, Inc.

2.6 Flashing

Flashing is needed in areas where water is likely to penetrate the roof system. Common areas for flashing include a pipe or vent penetration, the point where a roof meets a wall or other vertical surface (e.g., a chimney) or a roof valley. A valley, defined as an area where two different planes of a roof intersect at a low point of the slope, is where one can expect water to shed and accumulate before flowing downward to the gutter. Valley flashing is required as valleys are most susceptible to water infiltration. The most common detail for valleys is open. The shingle material is cut short of the valley, leaving the flashing exposed. Closed is when there is an overlap of shingles from one roof slope over the valley. The shingles on the other roof slope are cut parallel with the valley. The woven valley is not common because it is not considered aesthetically pleasing and is less efficient to install.

Any vertical penetration through the roof requires flashing. Step flashing, or smaller pieces of metal flashing are folded over the connection point and "step" their way up the roof slope to allow protection from water running down the roof. Step flashing is always laid from the lower side of the slope so water runs over the seam and not into it. The detail, as shown in Figure 2.6.1, must meet a 4-inch minimumfor asphalt shingles. Other roof covering types may require larger dimensions. The step flashing is laid alternating with the shingles.

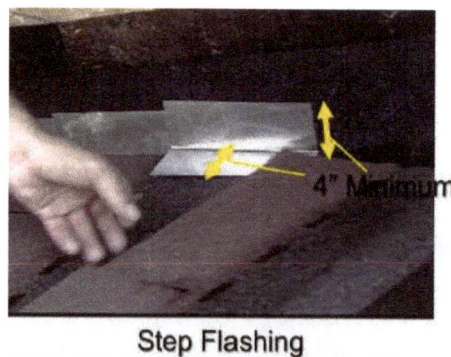

Figure 2.6.4– Step Flashing

Source: J.D. Miles & Sons, Inc.

2.7 Common Shingle Problems

Shingle problems can occur due to poor insulation, under-performing products, improper underlayments being used or fasteners not working correctly. Some common problems include curling, splitting (Figures 2.7.1, 2.7.2, and 2.7.3), cracking, granule loss (Figure 2.7.4) and shingle loss. These problems can shorten the lifespan of the shingles, resulting in water damage throughout the roof system. Granule loss will likely occur on shingles due to age, hail damage or abrasion (Figure 2.7.4). Granules are important to preserve because they provide shingles protection from water and from sunlight.

Figure 2.7.1– Curling and Splitting
Source: J.D. Miles & Sons, Inc.

Figure 2.7.2– Splitting
Source: J.D. Miles & Sons, Inc.

Figure 2.7.3 – Curling
Source: J.D. Miles & Sons, Inc.

Figure 2.7.4 – Granule Loss
Source: J.D. Miles & Sons, Inc.

Shingle loss happens mainly due to wind and/or poor application (Figure 2.7.5, Figure 2.7.6). Over time, with exposure to weather, shingle become brittle and can be damaged more easily. For high wind areas, six nail or four nail pattern is used to avoid shingle loss.

Figure 2.7.5– Poor Application of Shingles
Source: J.D. Miles & Sons, Inc.

Figure 2.7.6– Example of Shingle Loss
Source: J.D. Miles & Sons, Inc.

2.8 Tile and Wood Roofs

Roof tiles can be made from either clay (a natural element found in the earth) or concrete. There are several advantages to tile roofs: they have high fire ratings, indicating high resistance to fire spread; they have low maintenance requirements; they have pleasing aesthetics; and they have a long life expectancy. There are also a few disadvantages: they are extremely heavy and they are relatively expensive.

Figure 2.8.1– Clay Tile
Source: J.D. Miles & Sons, Inc.

Wood roofs are offered as either wood shingles or wood shakes (Figures 2.8.2 and 2.8.3). The advantages of wood include its natural beauty, its durability, its longevity, its resistance to the elements, its insulation value and its lightweight. Shingles are sawn on both sides and are thinner at the butt end when compared to shakes. Shingles are smoother on the surface and give a more uniform, flat appearance. Wood shakes are split during manufacturing, which allows for a variation in the thickness and texture of the final product. They have a rough appearance compared to wood shingles.

A few disadvantages include maintenance requirements, flammability (if untreated), cost, and the material's dimensional stability when exposed to changing moisture levels and temperature.

Figure 2.8.2 – Wood Shingles
Source: J.D. Miles & Sons, Inc.

Figure 2.8.3 – Wood Shakes
Source: Chris RubberDragon (CC Attribution)
https://www.flickr.com/photos/rubberdragon/9319035995

2.9 Tile Shingle Profiles

The National Roofing Contractors Association (NRCA) identifies different profiles that are widely recognized (Figure 2.9.1). The most common types are interlocking, flat and barrel.

Figure 2.9.2– Barrel Profile
Source: Pexel: Julia Vok (Creative Commons)

Figure 2.9.3– Flat Profile
Source: Ludowici
https://ludowici.com/products/roof-tile/

Figure 2.9.4– Interlocking Profile
Source: Ludowici
https://ludowici.com/products/roof-tile/

CHAPTER

03

Bituminous Low-Slope Roofing

With content donated by
Bruce McCrory
Nations Roof

3.1 Types of Low-Slope Roofing Systems

There are three major types of low-sloped roof systems: bituminous multi-ply, single-ply and liquid-applied, as outlined in Figure 3.1.1. Metal roofs are also gaining popularity and will be discussed later in this Manual.

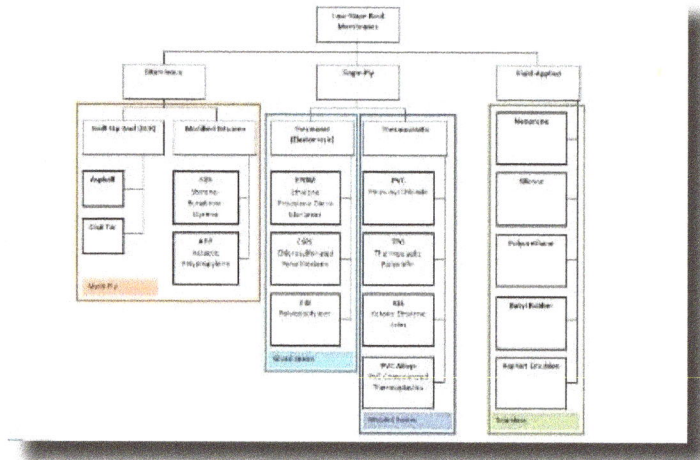

Figure 3.1.1–Types of Roofing Systems
Source: Nations Roof

Bituminous multi-ply roof systems are composed of multiple layers of roofing felts embedded in asphalt and often also include a cap sheet or top layer. They are the oldest and first, type of roofing system used in the U.S.

Single-ply roof systems provide a waterproofing layer using a single sheet and are the most common type of roof systems used in the U.S. today. Single-ply systems can be mechanically attached, ballasted or adhered depending on the project location and requirements.

Liquid-applied roof systems are applied in liquid form and cure to create a seamless waterproof membrane with no seams.

Metal roofs are composed of either structural or architectural panels. They are discussed in a later chapter.

In some parts of the country, sprayed polyurethane foam roofs perform successfully and can be used when other low-slope systems may not work, such as on domes. In this section, built-up roofs and modified bitumen roofs will be discussed.

3.2 Built-Up Roofs

A built-up roof (BUR) is a continuous, semi-flexible roof system consisting of multiple plies of saturated felts, coated felts, fabrics or mats assembled in place with alternating layers of bitumen (usually asphalt) and surfaced with mineral aggregate, bituminous materials, a liquid-applied coating or a granule-surfaced cap sheet (Figure 3.2.1). As its name suggests, this type of roof is built up on the job site.

Figure 3.2.1 –Built-up Roof (BUR)
Source: National Roofing Contractors Association

There are two main components of built-up roofing- ply sheet and bitumen. The ply sheet is either made up of matte or fiberglass or felt. Bitumen is another word used for asphalt, and there are different types of bitumen. The thicker the ply sheet, the more protection offered and the higher the cost. An example of type 3 and type 4 ply is shown in Figure 3.2.2

An important consideration for asphalt is its softening point because as asphalt is heated, it will soften into a liquid. The roofing specification should indicate the softening point for each type of asphalt to be used.

Exposure, side lap, head lap, and end lap are all important terms to know and affect the overall durability of the roof system. A diagrammatic representation has been shown in Figure 3.2.3 below.

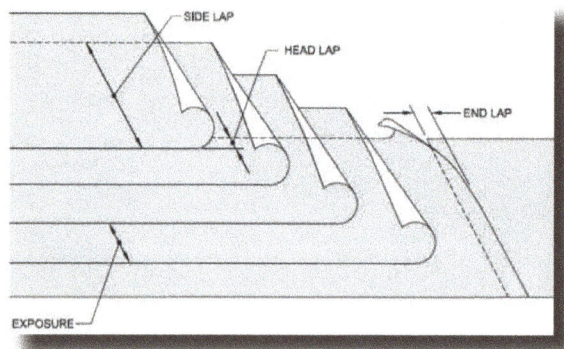

Figure 3.2.3 – Exposure, Side Lap, Head Lap, and End Lap

Figures 3.2.4 and 3.2.5 show different configurations for assembling built-up roof systems. The best way to identify a two-ply configuration is to draw a line at any given point in the roof system; then, the installer will pass through two layers. This is an uncommon system as it does not provide enough protection as a three or four-ply system. The exposure is less on a four-ply system compared to a three-ply system. Note that the cap sheet is not considered a ply. The cap sheet is only used for the surfacing material.

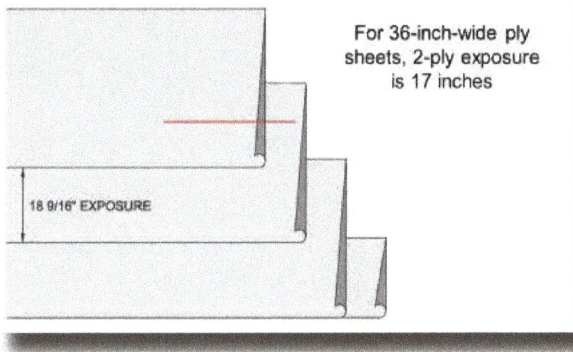

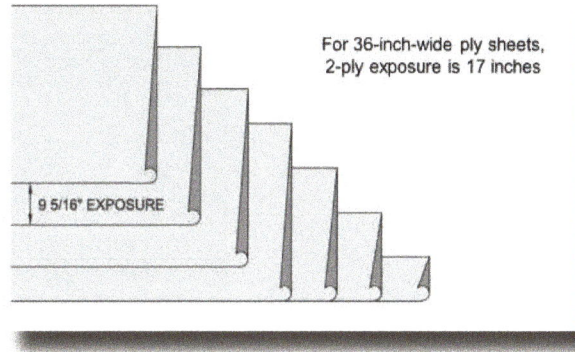

Figure 3.2.4 – Two-ply configuration Water Shedding Covering
Source: National Roofing Contractors Association

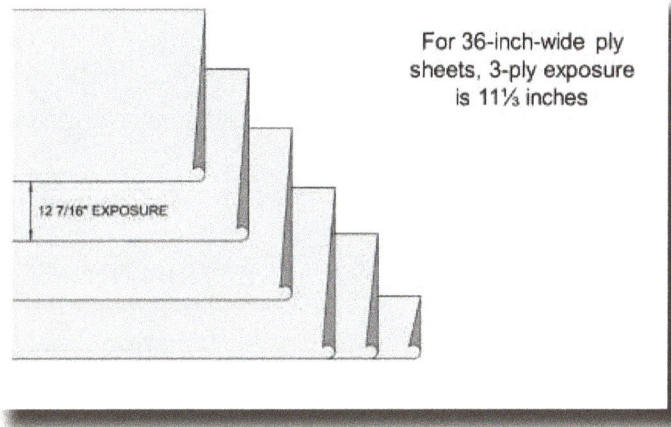

Figure 3.2.5 – Exposure, Side Lap, Head Lap, and End Lap
Source: National Roofing Contractors Association

3.3 Polymer Modified Bitumen

Another popular roof system is called modified bitumen. As its name suggests, it uses a bitumen that is modified by including one or more polymers (usually atactic polypropylene or styrene butadiene styrene) reinforced with various types of mats or films, and sometimes surfaced with films, oils or mineral granules. The big difference between built-up roof and modified bitumen roof systems is that the modified bitumen system is assembled in the factory and avoids extra work on the job site. (See Figure 3.3.1)

Fig 3.3.1 – Polymer-Modified Bitumen Membrane
Source: National Roofing Contractors Association

3.4 Surfacings for Built-Up Roof Systems

There are three types of surfacings for built-up roof systems:

A flood coat of asphalt with aggregate - The aggregates will adhere to the roof membrane and will help protect the roof from the sun (Figure 3.4.1).

Liquid-applied coatings - They come in buckets and are applied with a roller onto the BUR system. The most common type of liquid-applied coating is aluminum (Figure 3.4.2).

Cap sheets, which are similar to ply sheets but are made for surfacing and have granules embedded in them. They are installed with hot asphalt. (Figure 3.4.3).

Figure 3.4.1 - Aggregate Surfaced BUR
Source: National Roofing Contractors Association

Figure 3.4.2- Liquid-applied Coating
Source: National Roofing Contractors Association

3.5 Application Methods for BUR

The application of built-up roofs requires equipment to heat and apply asphalt as part of the system. The equipment used for BUR installation includes liquid asphalt tankers (Figure 3.5.1),

Heating kettles (Figure 3.5.2), and mechanical "felt layers," which are used to roll out the ply felts while simultaneously applying hot asphalt. Ply felts can also be applied using a mop, and equipment is also needed for flood coat and aggregate application. Installing a BUR system can be labor-intensive.

Figure 3.5.1 - Liquid Asphalt Tanker
Source: National Roofing Contractors Association

Figure 3.5.2- Heating Kettle
Source: National Roofing Contractors Association

Figures 3.5.3 through 3.5.6 show different methods of BUR application.

Figure 3.5.3- Rooftop Bitumen Handling
Source: National Roofing Contractors Association

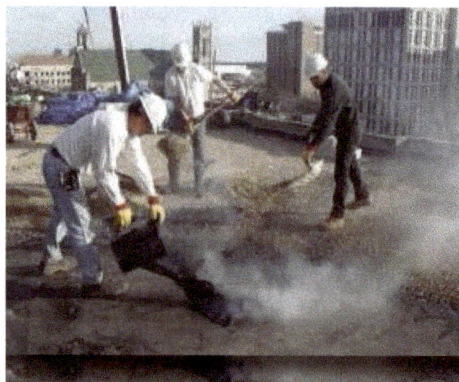

Figure 3.5.4 - Mop Application
Source: National Roofing Contractors Association

Figure 3.5.5- Mechanical Felt Layer Application
Source: National Roofing Contractors Association

Figure 3.5.6- Flood Coat and Aggregate Application
Source: National Roofing Contractors Association

The mop application process is a two-man job. One person rolls out the felt layers while the other person mops the asphalt behind him, as shown in Figure 3.5.4 above.

3.6 Modified Bitumen

The application of modified bitumen roof systems is similar to the application of BUR systems. Modified bitumen sheets are composed of polymer-modified bitumen (asphalt), reinforcement (usually fiberglass/polyester mats) and surfacing. The sheets can be unsurfaced, or surfaced with granules or foil. Unlike BUR systems, modified bitumen sheets are assembled in a factory. A common installation of modified bitumen roof systems uses a "base sheet," which is applied

on the roof deck, and a "cap sheet," which is applied over the base sheet and includes a surfacing material.

3.7 Application of Modified Bitumen Roof Systems

There are four basic ways to apply a modified bitumen roof system:

1. Torch application
2. Mop application
3. Cold adhesive application
4. Self-adhering application

Figures 3.7.1 through 3.7.4 show the different types of application. Note that in Figure 3.7.1, there is a roll and a torch attached to the equipment. The torch blows fire onto the roofing membrane to adhere the roll of modified bitumen on the roof system. The mop application is similar to that of BUR systems. For cold application, Figure 3.7.3 shows a machine called TCIV that lays cold adhesive down. The adhesive is used to attach modified bitumen to the deck system. This method is preferred because it does not require heating. Self-adhering application is not a preferred method of application because there could be some points where the sheet might not adhere fully with the adhering system. If there is heavy wind, this application is not recommended.

Figure 3.7.5 shows the variety of modified bitumen configurations. 3.7.6 shows a common cap sheet layout.

Figure 3.7.1– Torch Application
Source: National Roofing Contractors Association

Figure 3.7.2– Mop Application
Source: National Roofing Contractors Association

Figure 3.7.3– Cold Adhesive Application
Source: National Roofing Contractors Association

Figure 3.7.4– Self-adhesive Application
Source: National Roofing Contractors Association

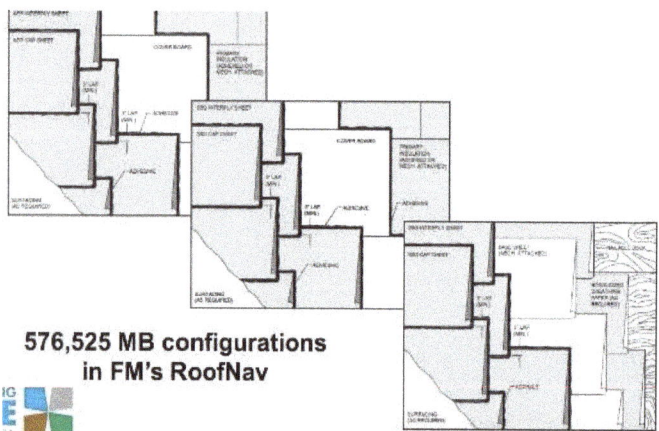

Figure 3.7.5– Modified Bitumen Membrane Configurations
Source: National Roofing Contractors Association

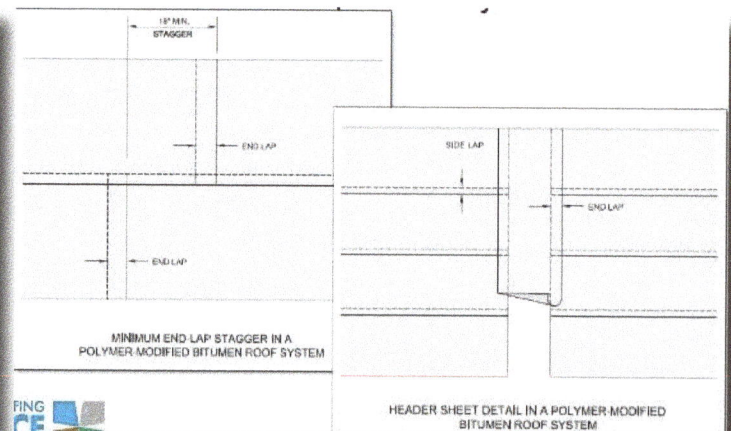

Figure 3.7.6–Modified Bitumen Cap Sheet Layout
Source: National Roofing Contractors Association

It should be noted that torches must be used with care. The National Roofing Contractors Association has developed a certification program for torch application (CERTA), and some insurance companies require workers using torches to be certified.

There are three different types of surfacings for modified bitumen roof systems:

1. Unsurfaced (where the sheet is smooth or has embedded granules)

2. Liquid-applied coatings

3. Flood coat (of asphalt) and aggregate

3.8 Roofing Details

For modified bitumen roof systems, like for all roof systems, it is important that flashing is done at the parapet wall, wall edges, or at any penetration such as drains, pipes, vents and mechanical units. Figure 3.8.1 shows flashing details at the parapet wall. Flashing is done to prevent water from entering any openings and cracks in the roof system.

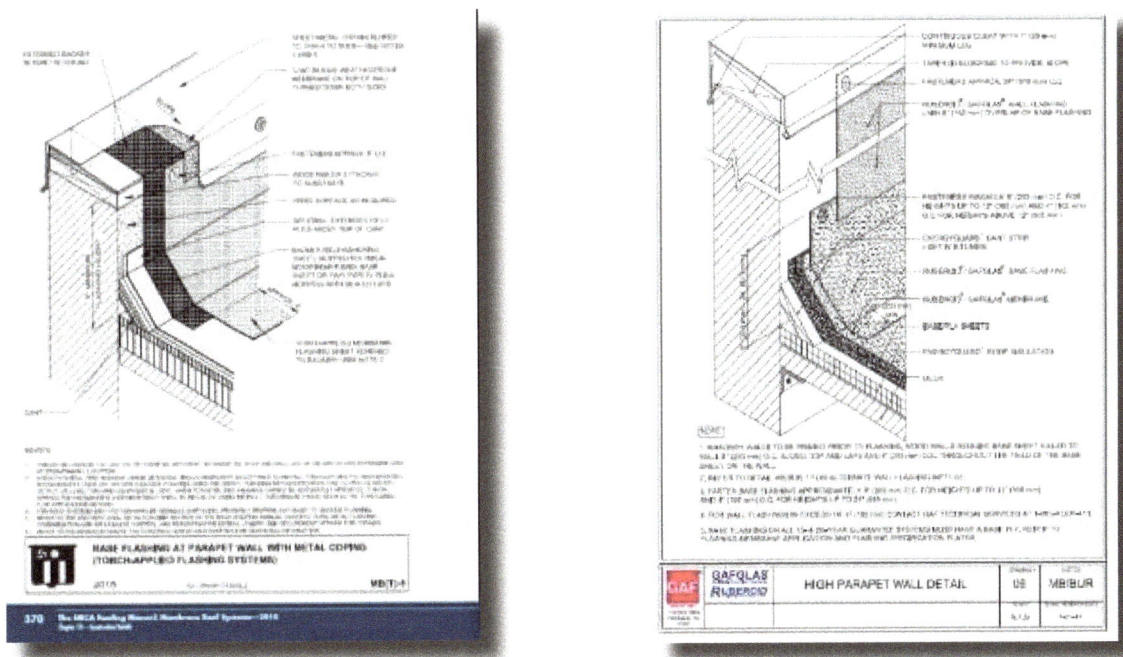

Figure 3.8.1- Flashing Details at Parapet Wall
Source: National Roofing Contractors Association

CHAPTER
04

Metal Roofing Systems

with content donated by David Landis
Petersen Aluminum Corporation

4.1 Structural vs. Architectural Metal Panels

There are two basic kinds of metal roof systems: structural, which is a "single-ply"-like membrane, or architectural, used as a decorative element (See Figures 4.1.1 and 4.1.2).

A roof with a pitch from ¼:12 to 3:12 requires a structural panel that is essentially watertight. These panels are called hydrostatic panels (Figure 4.1.3). Applications of 3:12 and above typically include panels that are hydrokinetic or water-shedding (Figure 4.1.4).

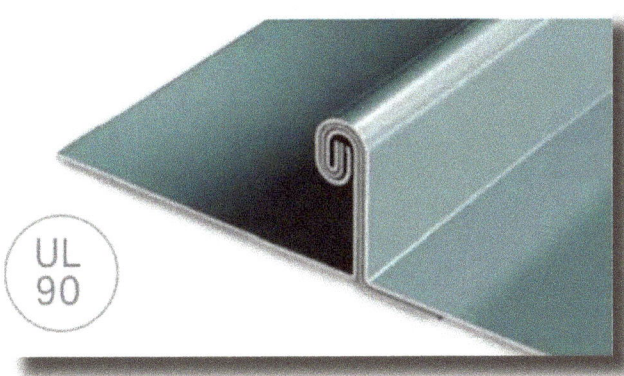

Figure 4.1.1– Double Lock Architectural Panel
Source: PAC- CLAD
https://www.pac-clad.com/products/metal-roofing/tite-loc-plus/

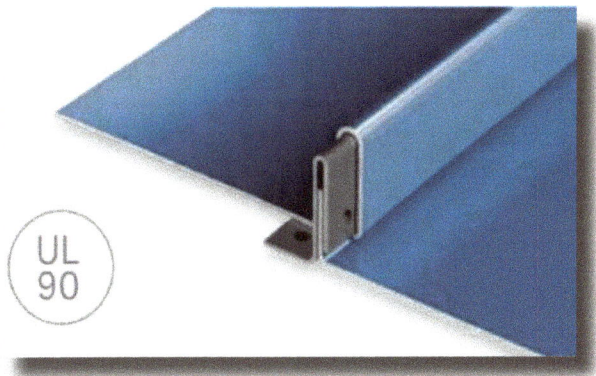

Figure 4.1.2– Architectural Panel
Source: PAC- CLAD
https://www.pac-clad.com/products/metal-roofing/redi-roof/

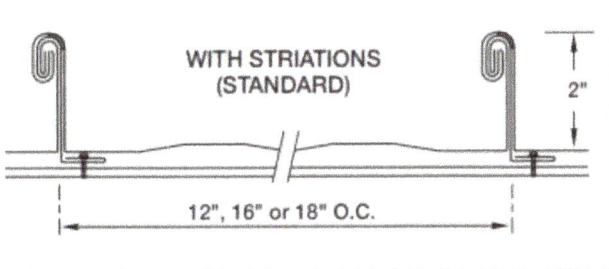

Figure 4.1.3– Hydrostatic Panels
Source: PAC- CLAD
https://www.pac-clad.com/products/metal-roofing/tite-loc-plus/

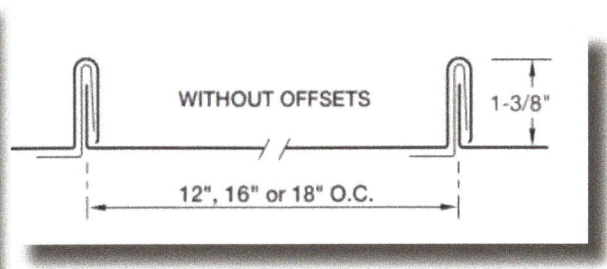

Figure 4.1.4– Hydrokinetic Panels
Source: PAC- CLAD
https://www.pac-clad.com/products/metal-roofing/tite-loc-plus/

4.2 Structural Panels

Structural panels are suitable for slopes as low as ¼:12. They have been designed to resist the passage of water under hydrostatic pressure. These panels should be rated by the Underwriters Laboratory (UL) for wind resistance (UL-90 or better) and include a floating clip to allow for expansion and contraction.

It should be noted that steel expands 1 inch per 100 lineal feet with a temperature change of 100 degrees F, and aluminum will expand 2 inches per 100 lineal feet with the same temperature change. The initial cost of structural panels is typically greater than single-ply roof systems

Structural panels are typically used in combination with batt insulation and may also require a thermal barrier such as a wood blocker. The overall life-cycle cost of these panels may be significantly lower than more traditional roof systems due to lower maintenance costs and a greater life expectancy. Maintenance typically includes cleaning and inspecting gutters annually. Typical applications include schools, industrial facilities, shopping centers, warehouses and prisons. Structural panel roof systems are generally not recommended for roofs with a vast number of penetrations and roofs exposed to caustic substances unless the panels are properly coated. Penetrations through metal panels need to be carefully planned so they can be appropriately detailed and flashed.

Structural metal panels can be field-seamed or snapped in place. Therefore, there are variations according to the type of panel that is specified. These variations include field seamed, field seamed long lengths, and no seaming required (snap in place).

Additionally, structural panels assembled over framing systems are frequently specified for reroofing. On these applications, a framing system is constructed over the original roof. In these situations, space between the original roof and the new metal roof may require the installation of a sprinkling system to satisfy code requirements. The seam from the bottom to the top is typically 2 to 3 1/2 inches tall, and the seams at the top require mechanical seaming.

Figure 4.2.4 shows a completed, structural panel roof.

Figure 4.2.1- Structural Panel
Source: AceRoofing (CC-Attribution)
https://commons.wikimedia.org/wiki/File:Radius_Panel.jpg

4.3 Architectural Panels

Architectural panels are typically preferred when the roof design is a visible, decorative element and an integral part oof the overall building aesthetics. Solid underlayment is typical with architectural panels. A minimum slope of 3:12 is typical, but is dependent on profile. Typical substrates include metal decking, 5/8" plywood, cover boards (oriented strand board laminated to rigid insulation) or Z-purlins in conjunction with rigid insulation. The span of Z-purlins may vary by panel and gauge.

Some architectural panels are suitable for application over structural framing in canopy applications. Virtually all architectural panels feature concealed fasteners and clips (Figure 4.3.1 and Figure 4.3.2). This is a big selling point to the owner. The fasteners are not visible in the finished application as they are under the panels are hidden by the overlapping panel.

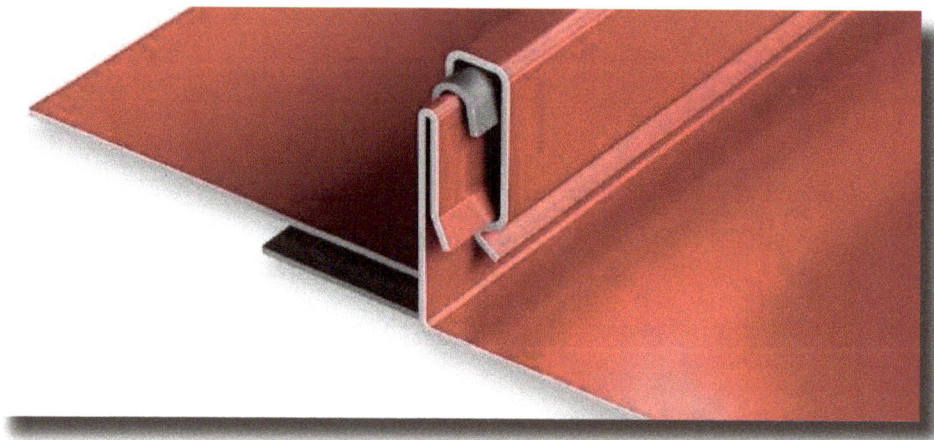

Figure 4.3.1 - Concealed Fasteners
Source: Petersen Aluminum Corporation

A "peel and stick" adhesive membrane is recommended for use under ALL standing seam metal roof systems, especially as more frequent and intensive weather events occur. All architectural barrel vault applications must have a peel-and-stick membrane as an underlayment. All underlayment should be applied in a shingle-like fashion horizontally from the eave to the ridge.

Aluminum is chosen over steel when the project is close to salt water, brackish water, or back-water bay, due to the superior resistance of the aluminum to saltwater corrosion (Figure 4.3.2).

Figure 4.3.2 - Aluminum Architectural Metal Roofing
Source: Petersen Aluminum Corporation

Figure 4.3.3 - Examples of Architectural Panels
Source: Petersen Aluminum Corporation

4.4 Components

There are different types of substrates for architectural panels, including galvanized steel, Galvalume, aluminum and copper. The panels are sent through a leveler and then rolled, as shown in Figure 4.4.1.

Figure 4.4.1–Leveler on Rollformer
Source: Petersen Aluminum Corporation

Galvanized steel has been successfully used for more than 75 years. Grade G-90 Galvanized is recommended for architectural applications. G-rating refers to the amount of zinc per square foot per side. The steel is available with or without a paint finish. It also has relatively low expansion and contraction.

Galvalume contains an alloy top coat which includes aluminum for corrosion resistance and zinc for corrosion/sacrificial properties. It has fewer sacrificial properties than a galvanized steel substrate but has better corrosion resistance. A warranty is available, usually for unpainted panels. Bare Galvalume should be kept out of direct contact with lead, copper, graphite, green wood, wet or treated wood, etc., as it may subject those materials to galvanic corrosion.

Aluminum has superior corrosion resistance, is flat and has good surface characteristics. Aluminum costs relatively more per square foot than some other products; has a high recycled content, is lightweight and has twice the expansion/contraction of steel. Because of its expansion/contraction characteristics, it may require modifications in the flashing design.

Copper is known for its corrosion resistance. It ages to a rich patina color that can be hastened by exposure to chemicals, as shown in Figures 4.4.2 and 4.4.3. An oxide wash should be factored into the design of a building with a copper roof to prevent staining on adjacent surfaces. Copper recently has had volatile pricing. It is a soft metal and requires a self-supporting underlayment. 16-ounce copper gauge is the most commonly used in architectural applications.

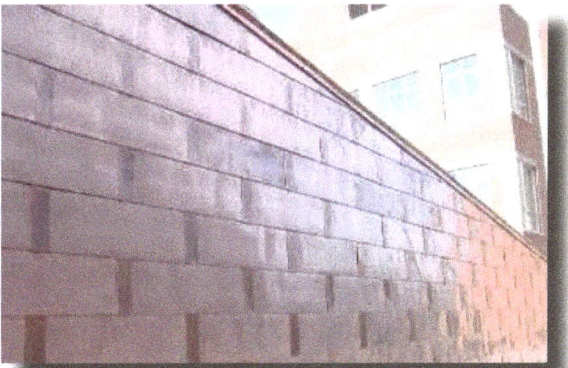

Figure 4.4.2– Copper in 10 days
Source: Petersen Aluminum Corporation

Figure 4.4.3– Copper in 2 years
Source: Petersen Aluminum Corporation

4.5 Metal Roof Finishes

The most prominent kind of finish for metal roof panels is called PVDF (for polyvinylidene fluoride). This coating is low friction and self-cleaning. Inherent qualities include weathering properties, ultraviolet resistance, formability, abrasion resistance and resistance to airborne pollutants. This coating is applied to G-90 Galvanized steel, Galvalume steel or aluminum and typically comes with at 30-year paint finish warranty for chalk, fade and adhesion for the metal substrates.

4.6 Potential Issues to Mitigate

One of the most common problems with metal roof systems is "oil canning," a visible, wavy distortion seen in the flat areas of metal roofs. There are several causes of oil canning, including an uneven substrate or deck, poorly adjusted rolling equipment, overdriven fasteners and/or expansion/contraction of the metal panel. It is also necessary for there to be a continuous use of roll along the seam; otherwise, there will be problems (Figure 4.6.1). It is essential that the contractor is aware of the use of appropriate rolling materials (Figure 4.6.2).

In Figure 4.6.2, the peel-and-stick underlayment does not cover the flashing, and the contractor has used galvanized nails rather than stainless steel screws to install the aluminum flashing.

Figure 4.6.1– Poor Seaming
Source: Petersen Aluminum Corporation

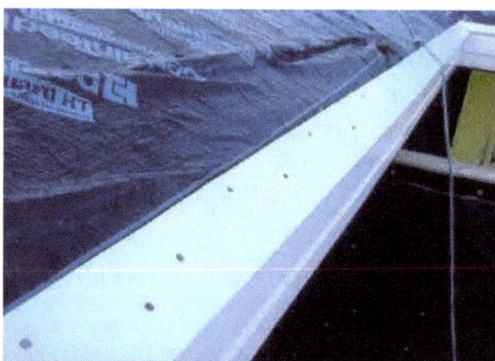

Figure 4.6.2– Incorrect Use of Roofing Materials
Source: Petersen Aluminum Corporation

CHAPTER
05

Low-Slope Single-Ply Roofing Systems
with content donated by
Mark Standifer
David Allor
OMG Roofing Products

5.1 Introduction

A single-ply roof system is a durable and efficient way to protect buildings of any size from the elements. They are a top choice among architects and engineers because of their cost-effectiveness and the clean look they give to the roof. With their multitude of colors, materials and installation methods, single-ply roof systems can waterproof a building without sacrificing the appearance or integrity of buildings.

5.2 Substrate and Decking

Before any roofing can take place, a roof must have a structurally sound substrate to attach to. Most commercial buildings use three main types of substrates: wood, concrete and metal   (Figures 5.2.1, 5.2.2 and 5.2.3). Types of wood decks include Plywood, OSB, and Dimensional Lumber. Wood decks are seen in smaller commercial construction. Concrete decks are lightweight and structural. Typically, concrete decks require moisture testing before any roof installation. Metal decks have different gauges and profiles. They are great for large spanning roof areas.

Figure 5.2.1– Wood Deck
Source: King of Texas Roofing, OMG Roofing Products

Figure 5.2.2– Metal Deck
Source: King of Texas Roofing, OMG Roofing Products

5.3 Insulation and Coverboard

To help insulate the building, thus reducing the energy necessary to heat and cool the building's interior, roof insulation is typically added to single—ply roof assemblies. Polyisocyanurate insulation is the most commonly used type of roof insulation; it is a thermoset plastic foam made up of polyurethane and methylene diphenyl diisocyanate.

Insulation is measured by its R-value. One inch of "polyiso" insulation will provide 5.7 units of R-value. The higher the R-value, the more energy efficient the building will be. Typically, large warehouses will require less R-Value, while conditioned spaces such as offices or restaurants will require more R-Value. The R-value is prescribed for different areas of the country as determined by the U.S. Department of Energy and local building code bodies.

An important note for laying out the insulation boards (typically 4'x8') is that most applications of "polyiso" are required to have two layers (5.3.1). The two layers should be installed with the joints of the boards staggered to prevent airflow and thermal transfer between the insulation.

Figure 5.3.1 - Insulation
Source: Ryo Chijiwa- Flickr (CC-Attribution)

5.4 Coverboards

Coverboards are mechanically fastened or adhered like polyisocyanurate. They are dense boards made of a number of different materials, such as silicone-treated gypsum and high-density polyisocyanurate composites. They give additional protection of the polyiso as well as additional impact resistance to hailstones or even heavy foot traffic for the roof system. Coverboards typically provide minimal, if any, R-Value. Certain manufacturers may also provide higher wind speed classifications or longer duration for the roof system's warranty when a coverboard is used.

5.5 Single-Ply Membranes

There are three main types of single-ply membranes that are used in the roofing industry today. Each membrane has its own unique traits and properties that make it ideal for certain conditions. The three most common membranes are TPO, PVC and EPDM.

TPO, which stands for thermoplastic polyolefin, is the most commonly used single-ply membrane today. It is easily installed in a variety of different methods. It can be mechanically fastened, adhered, or induction welded. It has numerous properties that make it ideal for most commercial roofing applications. It is available in a variety of colors, with white being the most popular for its reflective properties.

In terms of its structural capabilities, TPO membranes are manufactured with a scrim reinforcement fused between the two TPO membranes. The scrim provides rigidity and tensile strength for the membrane sheet, while the TPO polymers allow for the membrane to be heat welded to other sheets to create a continuous sheet of roofing.

PVC, or polyvinyl chloride, membranes look and act similarly to TPO in that they are rolled onto the roof and then heat welded. PVC has scrim reinforcement like TPO; however, the chemical composition of the membrane itself allows it to have a higher chemical resistance as well as a greater durability factor. Like TPO, PVC also has reflective capabilities due to the light color of the membrane. PVC can be adhered as well as mechanically fastened, and it is usually the recommended membrane for roofs that may be exposed to chemicals in industrial applications or grease on restaurant roofs.

EPDM (ethylene propylene diene monomer) is made of a synthetic rubber. Unlike TPO and EPDM, it can't be heat welded together to create a continuous sheet. Instead, it is adhered to other sheets using liquid adhesives or tape that is produced specifically by the manufacturer for that purpose. Because of this, EPDM seams need to be installed carefully.

There are three main methods of installation: mechanically attached, adhered or ballasted. Ballast for EPDM roofs is usually a layer of stone ballast, such as washed river rock, that is added to the top of the roof system to weigh it down and hold the roof system in place. EPDM is chosen sometimes because it usually costs less than TPO or PVC and because of its resistance to the cold and its fire resistance.

5.6 Single-Ply Details

Commercially available single-ply roofs have unique conditions that require attention at the perimeter, areas of drainage, and around HVAC units, air vents, solar panels, condensation lines and a multitude of other items. These areas are where the roof is most vulnerable, and the attention to detail needed by roofing contractors is what separates the best contractors from the rest and, subsequently, the best roof system from all the others.

Termination of the membrane at edges, walls and parapets are installed according to manufacturer recommendations and/or according to the NRCA standard details. When the membrane reaches an eave or edge of the roof, the membrane is sealed, or "flashed," with an edge metal of some kind. This might also require it to have a gutter to properly guide the water off of the roof. Similarly, when the membrane reaches a wall or parapet, it is typically flashed with a termination bar fastened to the wall itself or installed "up and over" the wall to be terminated at the top of the parapet with a coping cap.

Water drainage is one of the roof's primary purposes. As previously mentioned, roof gutters can be used at roof edges to catch the watershed from the roof and channel it into downspouts that guide the water to the ground. Roof drains (Figure 5.6.1 and Figure 5.6.2) are another way to channel water into pipes and drainage lines. The last method is to use roof scuppers, which are rectangular openings along the perimeter wall of a building that allow the water to leave. When using drains and scuppers, it is imperative to have a primary drain or scupper and a method for overflow in the event of an intense amount of rain.

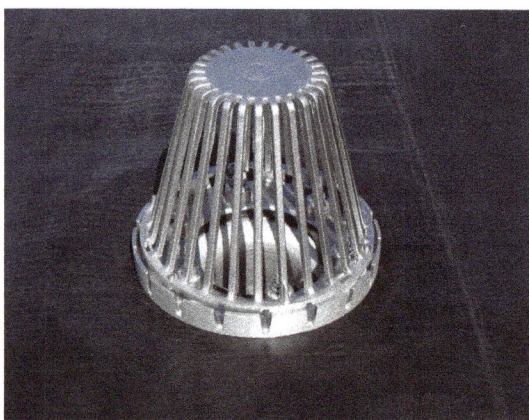

Figure 5.6.1– Roof Drain on an EPDM Roof

Source: OMG Roofing Products

Figure 5.6.2– Roof Drain on a TPO Roof

Source: OMG Roofing Products

5.7 Flashing Skylight Curbs

Roof curbs for HVAC units, skylights (Figure 5.7.1) and roof hatches also require special attention by the roofing contractor.

Curbs must be 8 inches above the roof membrane so that there is enough height for counter-flashings and to ensure the curb is protected from any rain or snow buildup.

Penetrations are another area of concern. Because of the cylindrical shape of pipes and conduits, preformed pipe boots (Figure 5.7.2), field fabricated wraps using reinforced membrane and pitch pans (Figure 5.7.3) are all used as methods to ensure watertight seals.

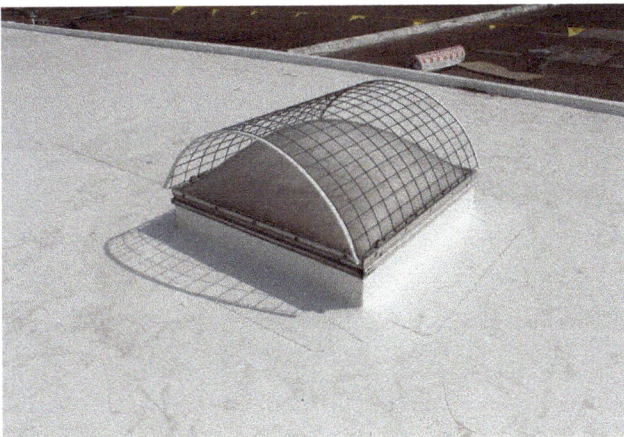

Figure 5.7.1- Flashing Skylight Curb
Source: OMG Roofing Products

Figure 5.7.2- Pipe Boot
Source: OMG Roofing Products

Figure 5.7.3- Pitch Pan Sealant
Source: OMG Roofing Products

CHAPTER
06

Liquid Applied, Waterproofing and Vegetative Roof Systems

with content donated by John Thomas

Siplast, Inc.

6.1 Introduction

Liquid-applied roofing includes products made from silicone, acrylic, polyurethane, polyether and others. There are many important applications for this type of roof system; liquid-applied products work very well on flashing details that have difficult transitions (Figure 6.1.1).

Figure 6.1.1 - Difficult Transitions
Source: Siplast, Inc.

Roof coatings are typically applied over existing roofing systems such as EPDM, PVC, TPO, Modified Bitumen, etc. Some of the advantages of using roof coatings include:

- High reflectivity
- Reducing the heat to a building
- Wearing surface
- Easier install
- Use of simple equipment such as rollers, brushes, squeegees
- Eliminates tear off
- Economical compared to roof tear-off

Silicone roof coatings are known best for UV stability and act better with ponding water. Silicones also have a broader application window because they are solvent-based and can be applied at broader temperature range. However, silicone coatings can be challenging to spray and might a slip-resistant surface to walk on. Acrylic roof coatings are cost-effective, and they also provide some UV resistance.

Acrylic roof coatings, being a water-based material, might not respond well to ponding water and can freeze at colder temperature. Hence, optimal outside temperature is needed to spray acrylic. Urethane roof coatings are commonly used for polyurethane foam insulation as well as over other roofing substrates. Urethane roof coatings are durable, flexible and provide reflectivity and UV stability.

It is critical that the existing roofing system is repaired and maintained before applying the roof coating. The roof coating is only as good as the roof they are covering. The roof coating can extend the life of an existing roof system and is a viable option available in the roofing market.

6.2 Waterproofing

There are many fundamental differences between waterproofing products, such as the ones discussed in this section, and roof coatings used for maintenance. Waterproofing products do not perform like paint and cannot be applied like paint. Liquid-applied roofing or waterproofing membranes have a long documented and successful performance history.

One popular product, PMMA – also commonly known as plexiglass, is widely used for road making, dental cement and industrial flooring. While it leaves a pungent odor until wholly cured, it is safe to use and poses no health risks.

Developed in the 1930s, the original PMMA formulation resulted in a rigid, exceptionally clear and durable material. In the 1970s, a resilient formulation was developed, allowing waterproofing and surfacing applications to benefit from PMMA. PMMA roofing, flashing and waterproofing is a multi-layered system consisting of material that is easy to mix, cures fast and is durable.

Another waterproofing system important in the roofing industry is the unreinforced waterproofing system, commonly found in parking decks (Figure 6.2.1).

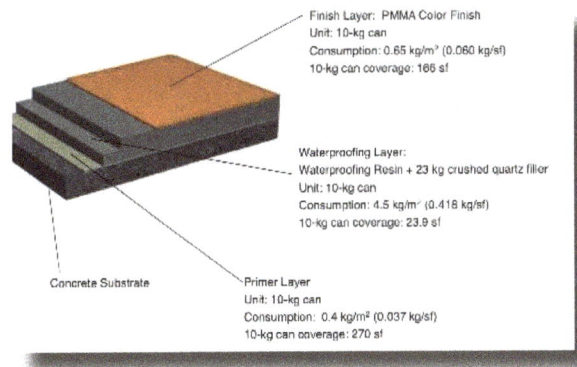

Finish Layer: PMMA Color Finish
Unit: 10-kg can
Consumption: 0.65 kg/m² (0.060 kg/sf)
10-kg can coverage: 166 sf

Waterproofing Layer:
Waterproofing Resin + 23 kg crushed quartz filler
Unit: 10-kg can
Consumption: 4.5 kg/m² (0.418 kg/sf)
10-kg can coverage: 23.9 sf

Concrete Substrate

Primer Layer
Unit: 10-kg can
Consumption: 0.4 kg/m² (0.037 kg/sf)
10-kg can coverage: 270 sf

Figure 6.2.1 – Unreinforced Waterproofing System
Source: Siplast, Inc.

The difference with this type of system is that the aggregate and quartz will be larger to accommodate the large, heavy vehicles that will be using the space. Because of cracking that can occur in the double Ts of concrete parking decks, the area gets permeated with a flexible material before the waterproofing material is applied. The curing time for parking decks is typically about six hours. The primer takes about 45 minutes; the waterproofing layer takes about two hours and the finish layer takes about three hours. A few examples of projects are shown in Figures 6.2.2 and 6.2.3 below.

Figure 6.2.2-Fire and Police Station in Burbank, California
Source: Siplast, Inc.

Figure 6.2.3 – Decorative Parking Deck
Source: siplast

6.3 Vegetative Roof Systems

A vegetative roof system, often called a "green roof," incorporates vegetation and a growing medium into the roof structure. Vegetative roof systems have many benefits, including reducing total energy loads, reducing peak energy loads, mitigating stormwater runoff and providing an aesthetic appeal. However, they are also expensive when compared to more traditional roof systems. One big advantage of vegetative roof systems is the reduction in temperature of the underlying waterproofing membrane, which will likely increase its longevity. A comparison between the temperatures of a vegetative roof and a standard roof system is shown in Figure 6.3.1.

Gravel Ballast Temperatures
1m above surface = 90°F
On surface = 122°F
On roof material = 118°F
In building = 92°F

Vegetated Roof Temperatures
1m above surface = 89°F
On surface = 98°F
On roof material = 96°F
In building = 83°F

Figure 6.3.1 – Green Roof Thermal Study
Source: Siplast, Inc.

Vegetative roofs are installed with a 4-inch growing medium underneath, with a root barrier provided (Figure 6.3.2). The pre-grown vegetated mat is grown in a field, rolled up and transported to the site, as shown in Figure 6.3.3. An aggregate-like material that absorbs moisture is grown in the field and rolled up for delivery to the roof. In this case, the building owner doesn't need to look for dirt for the vegetation to grow on the roof; sedum is part of the system. This technique ensures that the roof looks green immediately after installation.

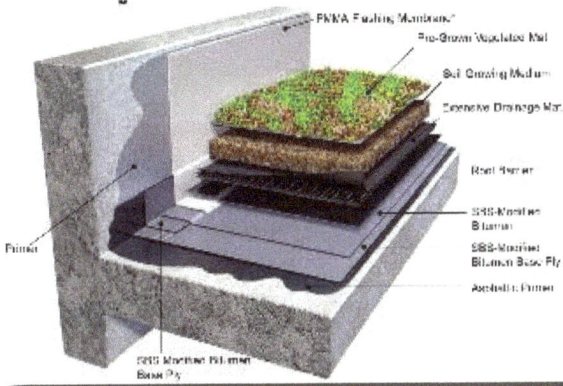

Figure 6.3.2- Pre-vegetated Extensive Roof System
Source: Siplast, Inc.

Figure 6.3.3- Rolled Aggregate Material
Source: Siplast, Inc.

In some cases, sprigs are installed by hand or mechanically inserted into the growing medium, as shown in Figure 6.3.4. Figure 6.3.5 shows installed sprigs at 18 months. Figure 6.3.6 shows installed sprigs at 36 months. Another possibility is the installation of "trays," which function similarly to individual potted plants (Figure 6.3.7).

Figure 6.3.4– Sprigs at Installation
Source: Siplast, Inc.

Figure 6.3.5– Sprigs at 18 months
Source: Siplast, Inc.

Figure 6.3.6– Sprigs at 36 months
Source: Siplast, Inc.

Figure 6.3.7–Trays
Source: Siplast, Inc.

The following are some examples of vegetative roofing products (Figures 6.3.8 to 6.3.12).

Figure 6.3.8– Ford River Rouge, Michigan
Source: Siplast, Inc.

Figure 6.3.9– Ford Roofing
Phase (Installation of Base Sheet)
Source: Siplast, Inc.

Figure 6.3.10– Javits Center
Source: Siplast, Inc.

Figure 6.3.11– Javits Center Base
Sheet Application
Source: Siplast, Inc.

Figure 6.3.12– Javits Center Roof Membrane Application
Source: Siplast, Inc.

6.4 Extensive and Intensive Roofs

Extensive vegetative roofs are ones with 4 inches of growing medium. They can house sedum, moss and grass. Intensive roofs, as shown in Figure 6.4.1, have a thicker layer of growing medium and can house sedum, moss, grass, large shrubs and even trees.

Figure 6.4.1 –Intensive Roofing System
Source: Siplast, Inc.

Another project, the NAVCC building in Culpepper, Virginia, involved the installation of a waterproofing/vegetative roof system (Figures 6.4.2 through 6.4.5).

Figure 6.4.2 Base Sheet Application
Source: Siplast, Inc.

Figure 6.4.3-Waterproofing Membrane
Source: Siplast, Inc.

Figure 6.4.4- Broadcasting of Vegetation
Source: Siplast, Inc.

Figure 6.4.5- Fully Grown Vegetation
Source: Siplast, Inc.

59

CHAPTER
07

Reading Roofing Drawings

with content donated by Andrew Christ
Cornell Roofing & Sheet Metal Company

7.1 Introduction

Drawings are technical documents usually produced by architects to communicate design, details and specifications. Engineers, architects, general contractors, subcontractors and consultants working in the construction industry should have the ability to read and understand drawings.

7.2 Subcontractor vs. General Contractor Drawings

The main difference between how a general contractor and a subcontractor look at a drawing is the level of detail. A general contractor is responsible for the entire project and will focus mostly on the details that are pertinent to the entire project. A subcontractor, on the other hand, has a specific scope of work. A subcontractor is responsible for ensuring that their bid includes everything from the drawings that is pertinent to their scope.

7.3 Drawings vs. Specifications

There are also specifications that accompany drawings. Drawings provide the drawings, dimensions, general details and so forth, while specifications give installers exact descriptions of how to install the product or system in accordance with manufacturers' guidelines.

7.4 Steps to Handle Drawings

Step 1

Typically, the owner sends the drawings to the general contractor, who shares them with the subcontractors as they work on estimating and scheduling. The subcontractor should note the bid due date and time because late bid submissions may lead to disqualification. The general contractor will also inform the subcontractors of any special requirements for the project.

A general contractor may have a template for the subcontractor to use. It is also important to see if the owner has requested alternates, for example, substituting a TPO roof system for a PVC roof system.

Step 2

The subcontractor should keep a job folder for every project; it should be where all documents related to the project are stored. The drawing index provided with the drawings provides sheet names with descriptions. It is important to go through all the pertinent documents, including the roofing plan, roofing details, section cuts and wall details to get a proper understanding of what is to be built.

Step 3

Extracting only the relevant documents from the entire set of drawings will help in reducing work to a manageable scale. Once the documents are extracted, they can be transferred to any estimating software platform.

Figure 7.4.1 show the details of a typical two-story school building. Note that a skylight is depicted on the roof, which means that details on finishing the skylight will need to be looked at. Detailed notes are provided throughout the drawing, such as the type of roof system, thickness of the coverboard, dimensions of the walkway pads and so on.

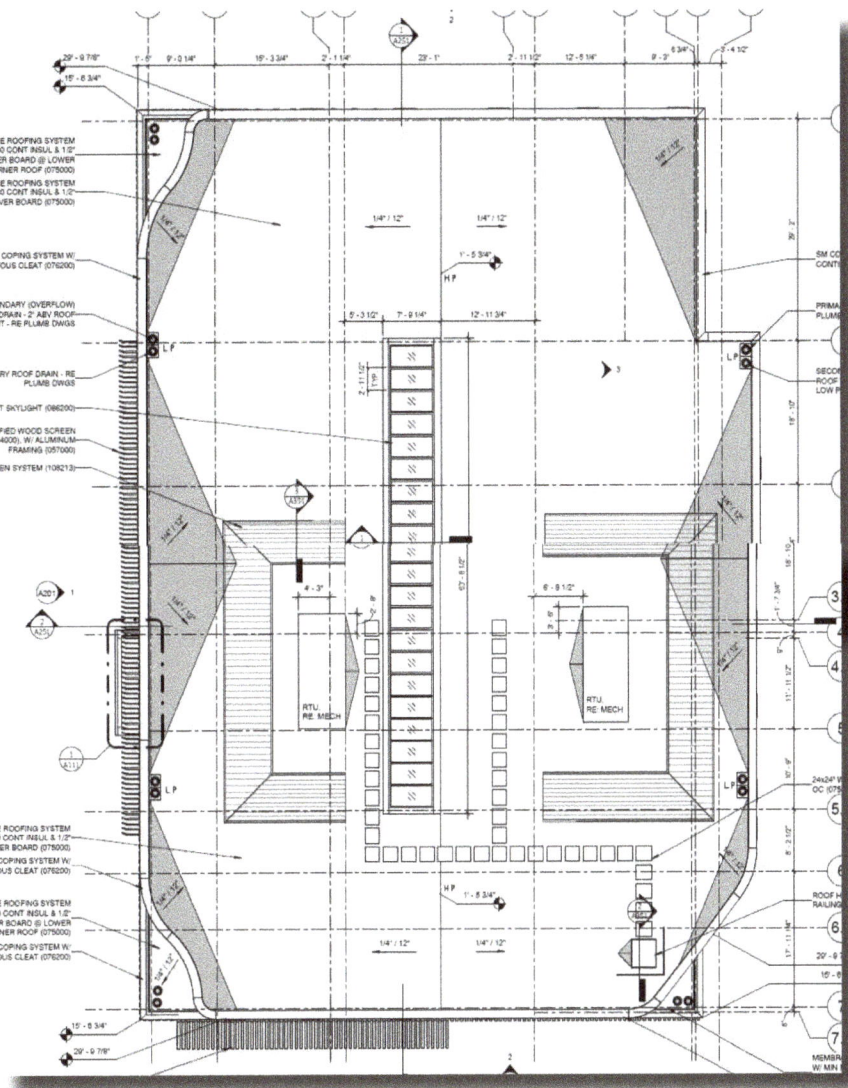

Figure 7.4.1– Roof Plan
Source: Cornell Roofing & Sheet Metal Company

The final step is to study the sheets that pertain to the scope of the roofing work. A common mistake in estimating is incorrect scaling of the drawings. Make sure the correct scale is used to get the right square footage. Any discrepancy observed in the drawings and specifications should be reported to the architect, general contractor and/or owner asking for clarification.

Figure 7.4.2 shows isometric sectional drawings that were provided for a project. They include information on how each corner of the roof will look and what material will be used.

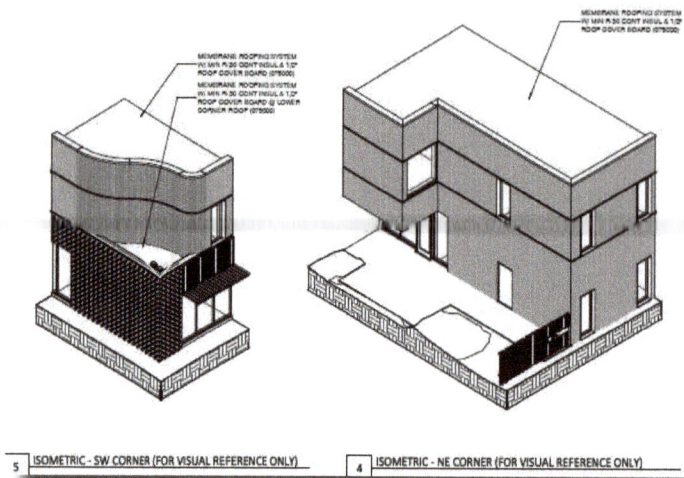

Figure 7.4.2 – Isometric Details
Source: Cornell Roofing & Sheet Metal Company

Building sections, as shown in Figure 7.4.3, provide important information, including the height of the roof. Knowing the height of the roof is important for determining how to get materials to the rooftop.

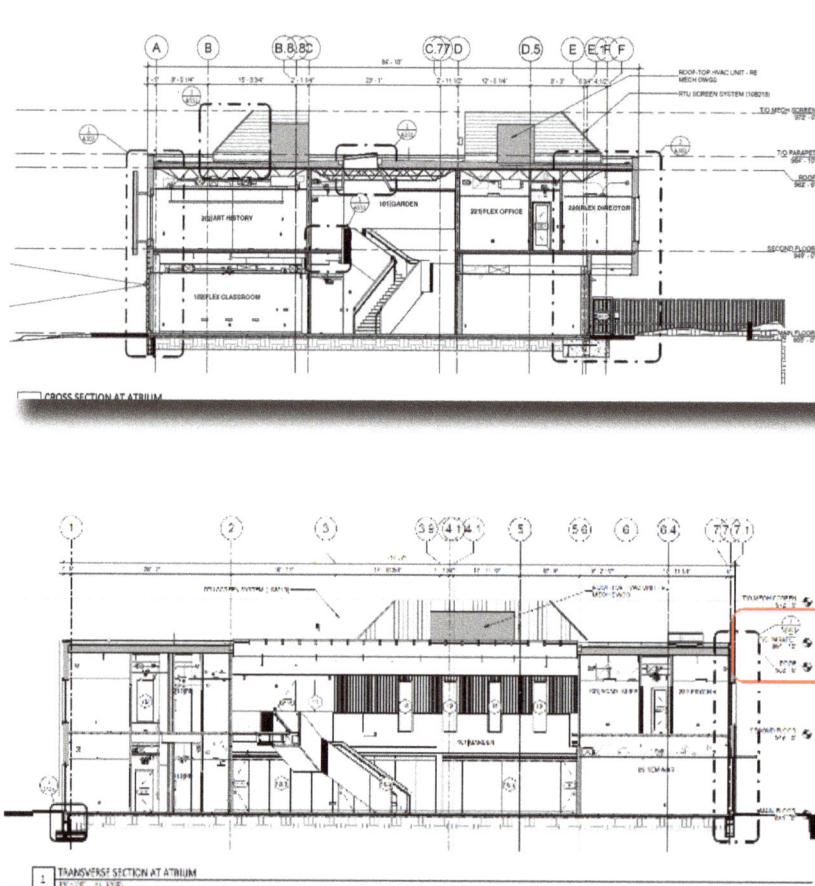

Figure 7.4.3 – Building Sections
Source: Cornell Roofing & Sheet Metal Company

Sectional details, as shown in Figure 7.4.4 below, provide information about the type of roof deck, thickness and R-Value of the insulation, coverboard, flashing configuration and roofing membrane

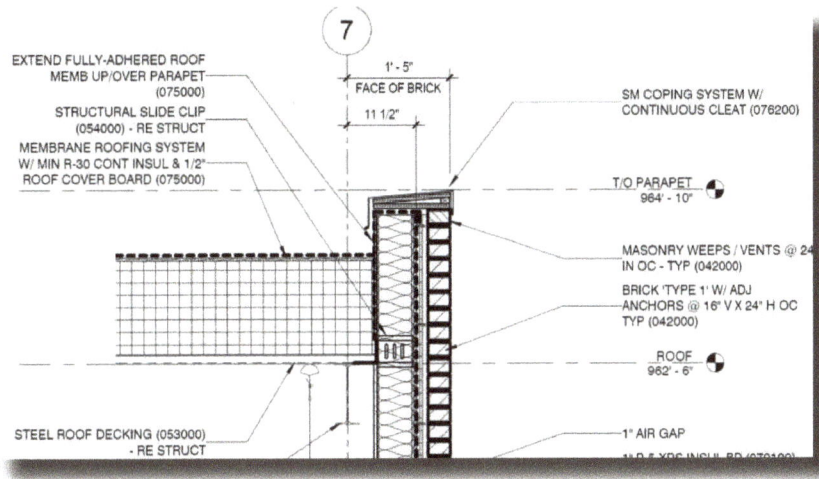

Figure 7.4.4 – Sectional Details
Source: Cornell Roofing & Sheet Metal Company

Enlarged exterior details, such as the one displayed in Figure 7.4.5, show how the parapet wall needs to be finished. This detail is applicable to all areas where the parapet wall is present..

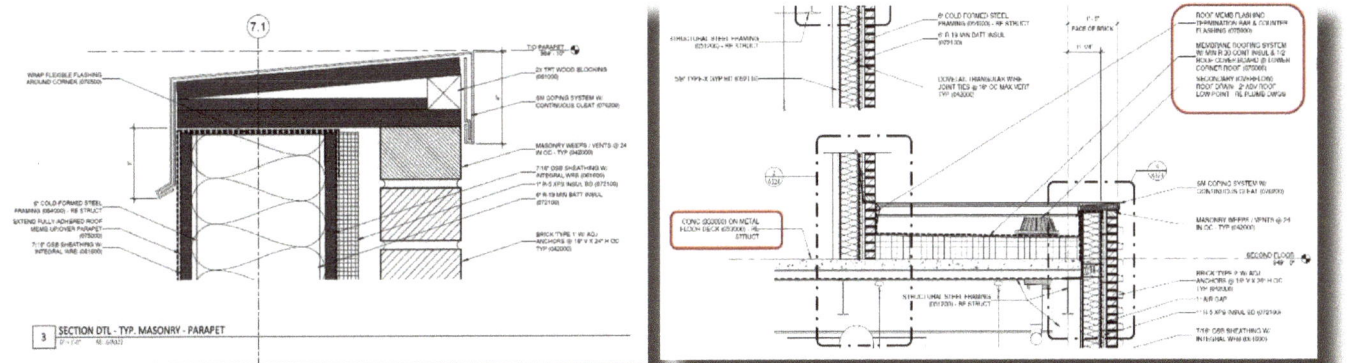

Figure 7.4.5 – Building Sections
Source: Cornell Roofing & Sheet Metal Company

It can also be helpful to review plan notes, as shown in Figure 7.4.6, provided along with the roof framing plan.

PLAN NOTES:

1. 1 1/2" - 22 GA TYPE B (WIDE RIB) PAINTED ROOF DECK. ATTACH TO SUPPORTING STEEL PER TYPICAL DETAILS T.O. ROOF DECK ELEV = 127'-7 1/2". REFER TO ARCH DRAWINGS FOR INSULATION SLOPES MIN 1/4" PER FOOT.

2. ALL SIMPLE BEAM CONNECTIONS SHALL BE PER DETAIL 1/S005 U.N.O.. DESIGN FOR LOADS INDICATED WHERE SHOWN. DESIGN FOR 10 KIP ASD, SERVICE LEVEL REACTION OF NO LOAD SHOWN. DESIGN MOMENT CONNECTIONS FOR LOADS SHOWN ON SHEET S008.

3. ALL PERIMETER ROOF EDGE ANGLES SHALL BE CONTINUOUS AND SPLICED IN ACCORDANCE WITH TYPICAL DETAILS.

4. 600S162-43 EXTERIOR WALL STUDS @ 16" O.C. UNLESS NOTED OTHERWISE.

5. SEE SHEETS S004, S005, & S006 FOR TYPICAL STEEL FRAMING DETAILS.

6. SEE SHEET S009 FOR TYPICAL DETAILS ON LIGHT GAGE CONSTRUCTION.

PLAN REFERENCE NOTES:

(A) L4x4x1/4 ANGLE FRAME AT ROOF HATCH PER TYPICAL DETAILS. REFER TO ARCH DRAWINGS FOR LOCATION.

(B) L4x4x3/8 ANGLE FRAME AT RTU'S PER TYPICAL DETAILS. COORDINATE LOCATION AND SIZE WITH ARCH DRAWINGS.

(C) JOIST SUPPLIER SHALL PROVIDE R TYPE JOIST EXTENSION DESIGNED FOR 50 PSF, ASD SERVICE LEVEL LOAD.

(D) JOIST SUPPLIER SHALL PROVIDE S TYPE JOIST EXTENSION DESIGNED FOR 50 PSF, ASD SERVICE LEVEL LOAD.

(E) DESIGN JOIST TOP CHORD AND ATTACHMENT FOR AN ADDITIONAL 8K SERVICE LEVEL AXIAL LOAD (ASD).

(F) DESIGN JOIST TOP CHORD AND ATTACHMENT FOR AN ADDITIONAL 2.5K SERVICE LEVEL AXIAL LOAD (ASD).

(G) CONTRACTOR SHALL COORDINATE HSS4x4x3/8 RAIL SUPPORT COLUMN LOCATIONS w/ ELEVATOR MANUFACTURER. REFER TO TYPICAL DETAILS FOR CONNECTION DETAILS.

(H) L4x4x1/4 ANGLE FRAME AT SKYLIGHT PER TYPICAL DETAILS. REFER TO ARCH DRAWINGS FOR LOCATION.

(I) PROVIDE HSS2x2x1/4 CROSSPIECE TO SUPPORT DECK @ ELEVATOR SHAFT. COORDINATE LOCATION w/ HOIST BEAM / ELEVATOR INSTALLATION REQUIREMENTS & ADJUST FRAMING SO THAT THE ROOF DECK SPAN NEVER EXCEEDS 6'-0". PROVIDE 1/2" BOLSTER PLATE @ EA SIDE & PROVIDE 3/16" FILLET WELD ALL AROUND TO ATTACH TO STEEL FRAMING.

(J) SCREEN ATTACHED TO STEEL FRAMING PER SCREEN SUPPLIER. ALL ROOF FRAMING DESIGNED FOR 1.8K (SERVICE LEVEL, ASD LOAD) IN ANY DIRECTION @ INDICATED LOCATIONS. COORDINATE FINAL SUPPORT LOCATIONS w/ SCREEN SUPPLIER.

(K) COORDINATE OPENING w/ ELEVATOR SUPPLIER.

Figure 7.4.6 – Plan Notes
Source: Cornell Roofing & Sheet Metal Company

Mechanical rooftop plans, as shown in Figure 7.4.7, indicate the placement of rooftop units. Plumbing, water and gas roof plans, as shown in Figure 7.4.8, provide details of roof drains, plumbing lines and penetrations on the roof.

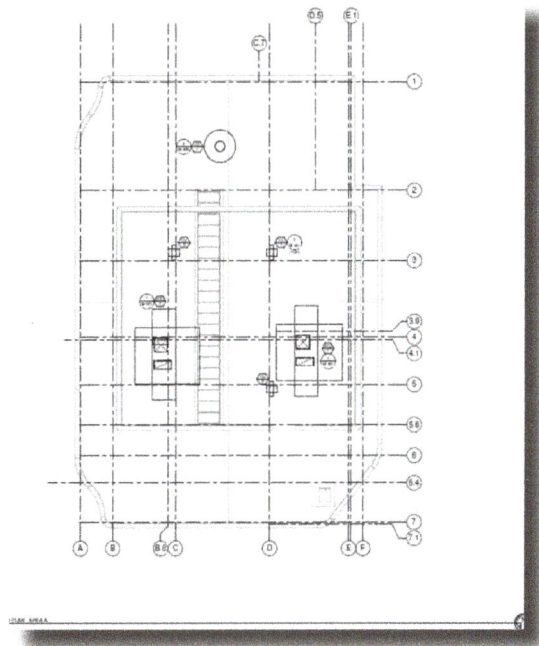

Figure 7.4.7 – Mechanical Roof Plan
Source: Cornell Roofing & Sheet Metal Company

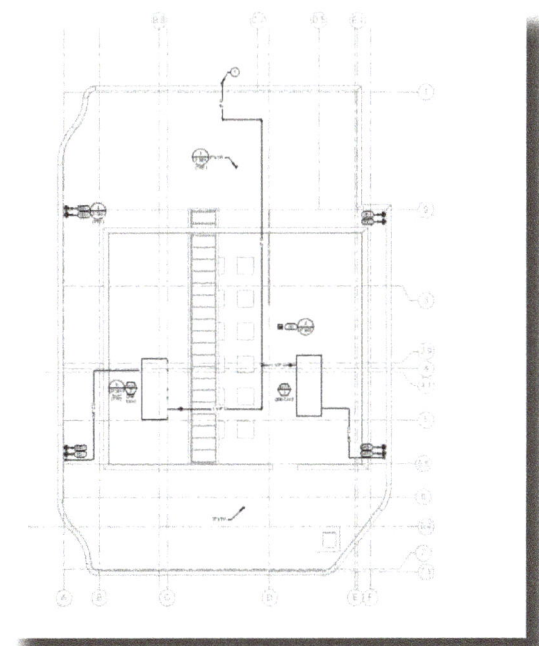

Figure 7.4.8 – Plumbing Roof Plan
Source: Cornell Roofing & Sheet Metal Company

Electrical roof plans, as shown in Figure 7.4.9, indicate electrical lines and connections for the HVAC equipment that will be placed on the rooftop.

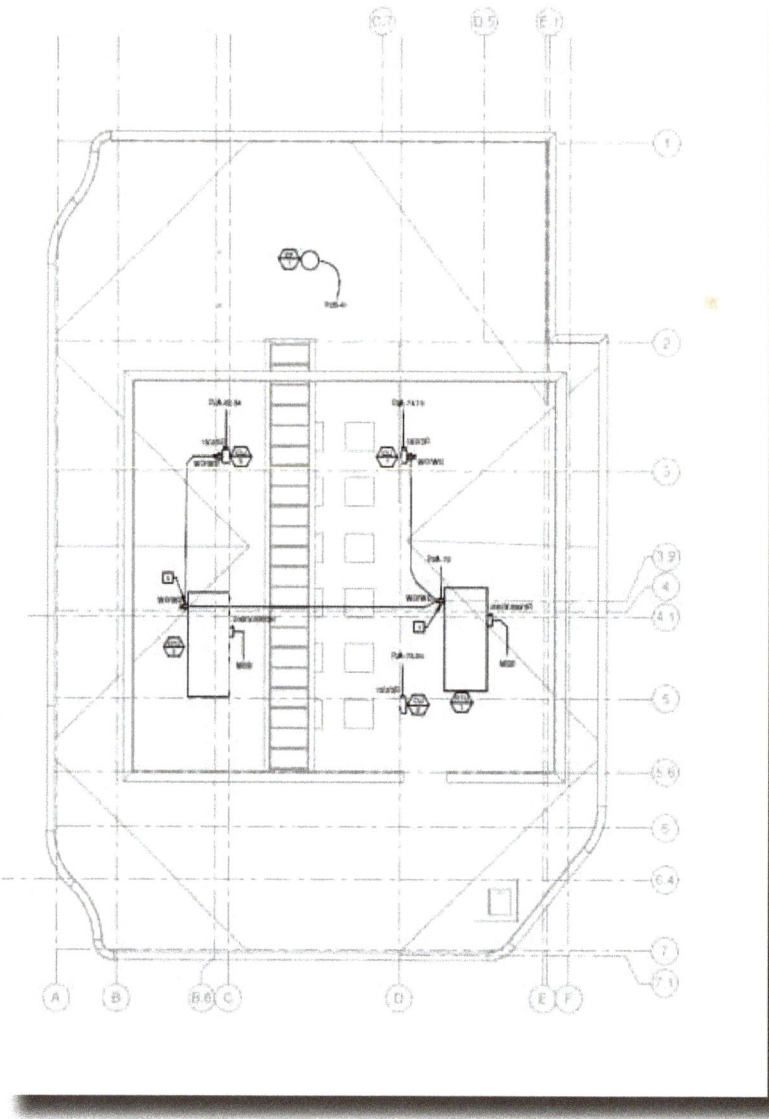

Figure 7.4.9 – Electrical Roof Plan
Source: Cornell Roofing & Sheet Metal Company

CHAPTER

08

Reading Roofing Specifications

with content donated by
Dan Jarratt- Piper Roofing
Justin Reeder- REI Engineers
Jeremiah Price- Soprema

8.1 Specifications

Specifications provide details about the type of roof system and details from the manufacturer about how to install the system. The specifications not only provide information about the roof system but also provide information about the qualifications of the contractor and the manufacturer.

Another way to ensure the roof system is being installed as specified is to conduct period inspections. From the owner's or general contractor's perspective, it is very critical to know that a qualified roofing contractor who knows the system is installing it exactly as specified.

Specifications are unique to each project and need not be memorized. It is important to know which sections to refer to for relevant information. All project activities and materials must meet the requirements of the specifications. Specifications and drawings should complement one another; if not, a request for information (RFI) should be sent to the owner.

Specifications for a job can come from AIA contract documents and are usually organized within the Construction Specifications Institute (CSI) MasterFormat standard.

8.2 AIA Documents

Documents published by the American Institute of Architects (AIA) and others listed below are an essential part of the specifications. These are standard forms that are commonly used in the roofing industry. They include:

- AIA A701 (Instructions to bidders)
- AIA 310 (Bid bond)

- AIA A101 (Standard form of agreement between owner and contractor)

- AIA A201 (General conditions of the contract)

Other state or governmental forms may be required for public jobs. Other owner specific requirements not found in the AIA documents may include tree/landscape protection requirements, existing property protection, work hours, etc.

8.3 CSI: MasterFormat

CSI MasterFormat is used by the construction industry to organize and communicate specifications across all phases of the project ranging from design, products, preconstruction activities, estimates, contracts, construction and facilities. Until 2004, the CSI MasterFormat consisted of sixteen (16) divisions. In 2004, MasterFormat increased to fifty (50) divisions to better encapsulate the specifications. Please visit csiresources.org to get a detailed description of specifications.

Information pertinent to roofing are often found in:

- Division 1: General requirements

- Division 6: Wood, Plastics and Composites

- Division 7: Thermal and Moisture Protection

Other divisions may also include relevant information depending on the scope of the project.

Division 1: General Requirements

Division 1 covers payment, administrative requirements and procedures for construction projects. The subcategories listed below provide information about permits, insurance, safety, environmental protection, etc.

Some of the information found in Division 1 includes:

- Summary of Work. These cover the basics required for bidding. The base bid is the estimated cost according to the specifications and drawings. Alternates are any additional options that the owner, architect or consultant might want a price for. For example, for a reroof project, the owner may want an asbestos survey to be conducted.

- Work Restrictions. These are any special conditions on the job site or any time restrictions for when the roofing contractor may work.

- Allowances. Allowances are for items that haven't been specified at the time of the awarding of the contract but are expected to be required for the project.

- Unit Prices

- Alternates. Often a contractor will prepare an additional bid along with the base bid that includes optional material substitutions or design features.

- Product Substitutions

- Contract Modification Procedures

- Payment Procedures

- Project Management and Coordination

- Submittal Procedures. Submittals assist in avoiding errors. The process involves providing specifications , samples, colors, etc. for all materials and shop drawings for any special installation requirements on the job. It is always a good idea fo get all the approvals on time to avoid schedule delays.

- Quality Requirements.

- References

- Temporary Facilities and Controls. These include fencing, scaffolding systems and other temporary fixtures that need to be included in the base bid. They may not seem to be an important part of specifications, but they can be a considerable cost and cannot be ignored.

- Cutting and Patching

- Cleaning and Waste Management

- Closeout Procedures. Closeout procedures refer to the documentation and activities performed during the completion of a project. Some of the steps that are a part of closeout procedures are:

    1. Punch list. During the final walk-through inspection, a list of all outstanding work items is made. The contractor is responsible for completing work on the punch list before project completion.

    2. Remove temporary facilities. Fences, portable toilets, material storage areas and other temporary facilities used for the project need to be removed once construction has been completed.

    3. Waiver/Releases. A waiver/release is a legal document that removes the contractor's liability for certain claims or obligations.

- Record documents/warranties. Record documents are all the project-related documents produced during the entire project such as design drawings, change orders, submittals and specifications. These need to be handed over to the owner to maintain and operate the building after construction. Warranties are provided by manufacturers, suppliers and contractors on materials, equipment and services provided during construction.

Division 6: Wood, Plastics and Composites

This section consists of specifications for products made of wood, plastic and composite materials. Some of the information in Division 6 relevant for roofing may include:

- Maintenance of wood, plastics and composites

- Rough carpentry

- Finish carpentry

- Exterior carpentry

- Architectural woodwork

- Structural plastics

- Composite fabrications

- Structural composites

Division 7: Thermal and Moisture Protection

Some of the information in Division 7 relevant to roofing, including:

- Maintenance procedures. Preparation for reroofing is one part of this subcategory that is very important. This part provides instructions for removing an existing roof and is not applicable to new construction.

- Damp proofing and waterproofing. This section provides information about the installation of systems and materials that protect the building from moisture. These materials include waterproofing membranes, drainage systems and coatings.

- Thermal protection. This subcategory provides information about roof and deck insulation. It includes information about R-Values, materials and applications. It is important to know the difference between the minimum R-value and the average R-value of the insulation. It is also important that the specifications comply with the local building codes. If the specifications require a minimum R-value of R-20, it means the minimum R-value should be R-20 or above at all locations. For tapered insulation systems, if the R-Value needs to be a minimum of R-20, then the insulation will need to have a greater R-Value at the center of the roof, to provide for an R-Value of 20 throughout the roof system.

- Steep-slope roofing. This section provides information about shingle, roof tiles and natural roof coverings.

- **Roofing and siding panels.** This section provides information about roof panels, wall panels, faced panels and siding.

- **Membrane roofing.** This section covers elastomeric membrane roofing, thermoplastic membrane roofing, liquid-applied roofing, etc.

- **Flashing and sheet metal.** This section covers guidelines for the installation of metal components for flashing and coping, as well as information about acceptable materials that can be used for coping, counter flashing, vent pipes, expansion joints, scuppers and son on.

- **Roof and wall specialties and accessories.** This section consists of information about roof hatches, scuppers, drains and rooftop equipment supports.

- **Fire and smoke protection.** This section provides information about smoke seals, fireproofing materials and their installation.

- **Joint protection.** This section provides information about sealants used for sealing joints and roof penetrations.

CHAPTER

09

Estimating I: Estimate Organization

with content donated by
Andrew Christ
Cornell Roofing & Sheet Metal Company

9.1 Introduction

A general contractor is responsible for the overall management of the project and has a general understanding of the scope of the project. General contractors, however, do not focus on project specifics and typically bring in specialty trade contractors, such as roofing contractors, for each component of the project.

A general contractor is primarily focused on the following tasks:

- Facilitating the bidding process by communicating and updating information necessary for bids to be submitted.
- Using specifications and plan take-offs as a guide for qualifying bids
- Identifying gaps in the scope of work to make the bidding process fair
- Evaluating and awarding bids

A subcontractor (roofing contractor, for purposes here) is responsible for developing a price for the work it will perform and is also responsible for the following tasks:

- Reviewing all project documents, including specifications and trade-specific sections
- Reviewing drawings for content, clarity and conflict.
- Developing and sending Requests for Information (RFIs) to the general contractor
- Finalizing an estimate by ensuring design and warranty requirements are met and complete systems are being proposed.

Roof estimation and general contractor estimation share many similarities but are also very different. The typical scenario for a new project involves the roofing contractor submitting an estimate to the general contractor.

However, for reroof projects or major repairs, the roofing contractor often works directly with the owner (or the owner's agent, such as a roof consultant). Roofing estimates include detailed material pricing for all the roof components, along with labor rates and profit margins.

9.2 Pre-Bid Meetings

Pre-bid meetings are usually held by owners about two weeks after specifications and drawings have been sent to the company or companies invited to bid on the project. The objective of the meeting is to give the contractors an overview of the project, its details, policies that are applicable to the project site and any owner-specific conditions that may exist. Contractors are also given the opportunity to ask for clarification on any of these topics.

A pre-bid meeting will help the contractor to evaluate the nature of the job, the type of workers required, staging areas and other logistical issues, and other items that may be required to meet the owner's expectations for the quality of the work.

Some pre-bid meetings, such as those for state or federal government projects, are mandatory. For these projects, contractors can only bid for the project if they were present at the pre-bid meeting.

Even in non-mandatory meetings, it is important for the contractors so they don't miss any information that will help them generate a proper bid.

9.3 Estimate Checklist

For a roof replacement project, it is imperative to conduct a site visit before determining the scope of work. A site visit gives a comprehensive overview of existing conditions, which drawings do not. While on a site visit, three critical tasks should be performed:

- Verify measurements. The onsite roof measurements must match those provided in the drawings.

- Take photographs of existing conditions. Existing conditions need to be well documented in the event of disputes and serve as a reference point after construction begins.

- Perform core cuts to determine the composition of the roof system. Core cuts provide important information about the existing roof, including what parts will need to be removed. That will enable a good estimate of the labor required to tear off all or parts of the existing roof. Core cuts also help determine whether the roof is structurally sloped or has tapered insulation and the thickness of the insulation.

9.4 Review of Specifications

The first step in reviewing drawings and specifications is identifying the type of roof system proposed. A specification consists of three parts with several subsections:

Part One: General specifications. This includes what is expected from the subcontractor, including submittal requirements, quality control requirements, existing project conditions and warranty requirements.

Part Two: Product Specifications. This part lists the products that need to be used for the roof system. Special attention needs to be paid to standards developed by the American Society for Testing and Materials (ASTM) and grade types.

Part Three: Execution Specifications. This part provides parameters and guidelines for the actual installation of the roof system.

9.5 Alternates and Addenda

Alternates include alternate pricing for optional scopes of work. Common alternates include:

A different roof system. The owner might request an alternate estimate for using a roof system other than the one provided in the specifications.

More robust system options or warranty durations. The owner might request an alternate estimate for a roof system with a longer life expectancy and longer warranty term.

Addenda are changes to the contractual scope of work added throughout the bidding process. Addenda drawings indicating these changes are sent to subcontractors, who need to incorporate them into the bid or risk having their bid rejected.

9.6 Building Codes

Each municipality has its own building regulations, which makes it essential to understand the codes that apply to every project. The roofing contractor must be sure to determine the frequency of inspections that are required, the cost of a building permit and whether there is a cost for a permit when the contractor works directly with the building owner.

9.7 Takeoffs

Takeoffs are intended to enable the contractor to determine the size and quantity of materials needed for a project. Because takeoffs form the foundation of an estimate, it is crucial to ensure that they are as accurate as possible. Developing an accurate takeoff becomes more manageable with the help of software and services currently on the market. Those include The Edge Roofing Estimating Software, On Center Takeoff, Square Takeoff and EagleView.

There are four principal elements of a takeoff:

1. Square footage of the field of the roof
2. Lineal footage and square footage of flashings, with special attention to flashing details provided in the drawings
3. The layout and amount of tapered insulation that is required. Because tapered insulation comes in different thicknesses, additional insulating layers may be required to meet the required R-value.
4. Verify that roof slopes match the slopes shown in the drawings.

9.8 Material Pricing

The first step in determining material pricing for the project is to make a detailed list of all materials and quantities required. This list should be sent to the local distributor or manufacturer for unit prices of each material on the list.

9.9 Takeoff Spreadsheet

A takeoff spreadsheet consolidates the pricing of materials, labor and extras. The spreadsheet provides a final estimate for the project and functions as a uniform bid sheet for the reference of other employees. The takeoff spreadsheet is used to determine the Schedule of Values (SOV), to track labor and production, and for work-in-progress (WIP) sheets.

9.10 Profit Margins

Gross profit refers to the amount remaining after direct job expenses, which mostly include material and labor costs. Net profit is the amount remaining after including indirect costs, such as fuel, rent, utilities, insurance and marketing. According to a study by the National Roofing Contractors Association (NRCA), a roofing company makes an average net profit of 2.8%.

9.11 Other Expenses to Keep in Mind

- Payment and performance bonds - A payment and performance bond typically accounts for 1 – 2.5% of the total bid amount. This bond is not always required but can be found in most projects.

- Permit fees - It is essential to determine if a building permit is required for the project and how much it costs. If a general contractor is involved in the project, they usually incur building permit costs.

- Taxes - Tax rates vary according to municipalities, and some projects may be exempted.

- Material handling and other logistics - The roofing contractor must take into account how the material will be delivered to the job site and loaded on the roof. In heavily populated urban areas, this may become a significant added expense.

9.12 Submitting the Bid

After receiving the Invitation to Bid from the general contractor or owner, it is crucial to note the bid time, date and location of the bid opening. The specifications will also come with a standard bid form that must be followed. Follow-up documents, such as affidavits and bid bonds, should be kept handy to be submitted along with the bid form.

CHAPTER
10

Estimating II: Considerations

with content donated by
Candace Klein
Klein Contracting Corporation

10.1 Equipment

Every roofing project will require its own set of equipment. Any equipment valued at $500 or more is classified as Capital Equipment. Small tools refer to any equipment valued at less than $500. Information on equipment pricing is available in the NRCA Equipment Cost Schedule from the National Roofing Contractors Association. Figure 10.1.1 is an example from the NRCA Equipment Cost Schedule showing hourly, daily, weekly and monthly rental rates for different types of equipment. For a project requiring lifting equipment such as a crane or forklift, it is best to contact a local crane or forklift representative for current pricing.

ROOF CUTTERS

DESCRIPTION	COST	MONTH	WEEK	DAY	HOUR
Hydra Saw	$5,926	$1,039	$364	$135	$16.62
Low-profile, mini-saw, 5 h.p.	$2,439	$428	$150	$56	$6.84
Single-blade, 9 h.p.	$3,023	$530	$185	$69	$8.48
Double-blade, 13 h.p.	$3,174	$556	$195	$72	$8.90
Double-blade, 20 h.p.	$6,312	$1,107	$387	$144	$17.70
Self-propelled hydraulic, double-Blade, 20 h.p.	$7,487	$1,313	$459	$171	$21
Self-propelled twin saw, 16 h.p.	$8,454	$1,482	$519	$193	$23.71
Saw with water sprayer and H.E.P.A. vacuum 9 h.p.	$7,044	$1,235	$432	$161	$19.76
*Add-on water spray system	$424	$74	$26	$10	$1.19

Figure 10.1.1 – NRCA Equipment Cost Schedule
Source: NRCA

The roofing contractor must plan for all of the equipment that will be used on a project and include pricing for it in the project estimate. Estimating software typically does not include equipment costs, so they must be tracked on a separate spreadsheet. If a project requires equipment not included in NRCA's schedule, it is advisable to contact a local renting company for the most current pricing.

10.2 General Conditions

General conditions include any other project requirements that do not fall into direct material or labor costs. These include:

- Overhead, including office costs, energy, utilities, administration, insurance, marketing, etc., which are not part of the project activities but are required to operate the business. Overhead is usually 10 to 15% of the project's total cost.
- Supervision. Some contractors include supervision as part of overhead; others calculate the hours of supervision necessary for each project.
- Capital equipment – valued at $500 or more.
- Rental equipment
- Supplies, such as paper, printers and computers
- Portable toilet facilities
- Shop drawings
- Field office
- Small tools – valued at less than $500
- Building permit

10.3 Roofing Crew Details

Generally, a roofing crew consists of a superintendent, foreman, journeymen, laborers and apprentices/helpers. This, of course, will vary by company. An average roofing crew is between five and ten workers.

Superintendents are authorized to supervise work on site; they also are responsible for meeting with the general contractor and/or owner and overseeing the progress of the work. A foreman is considered the leader of the work crew and is responsible for its management.

Journeymen and laborers work under the foreman and are experienced in roof installation. A helper assists the journeymen and laborers in tasks such as moving equipment and materials; they are typically new to the job.

Depending on the size and complexity of the project, there might only be one crew handling the entire scope of work, or there may be multiple crews on a project. Crew size and composition depend in large part on the project's schedule and completion date. It is not always true that more crews will result in faster completion of the project; crews must be able to maintain a level of productivity and not get in each other's way.

10.4 Labor Cost

Step 1 - Determine Quantities

Determine quantities, including the square footage of the field of the roof, the square footage of flashings, the amount of insulation required and accessories that are part of the specifications.

Step 2 - Apply Man-hour Costs to Labor

Every company will have man-hour rates available for different employees on the project. These will include the employees' wage, plus taxes, worker's compensation insurance, general liability insurance and fringe benefits.

Step 3 - Productivity Rate

Productivity rate refers to the net output produced in an hour of work. This is usually calculated as square foot per man hour, man hour per square foot, man hour each or man hour per linear foot. Every company uses its own method for making these determinations, which also vary by activity.

Step 4 - Total Man Hours

The quantities produced during Step 1 and the productivity rates determine the total man hours required for the project.

Step 5- Total Labor Costs

The total labor cost is determined by multiplying the total man hours with the average labor man-hour cost.

10.5 Calculating Total Man Hours

The following is an example showing how to calculate total man hours on a project. The project in this example is a 25,000 square foot TPO roof membrane over concrete and 3' high wall flashing for 350'. The job is to be done with one crew consisting of one foreman, one journeyman and three laborers/helpers.

From the scope of work, it can be determined that there is 25,000 SF of TPO Membrane and 3'height wall flashing for 350'.

The total man hours needed from this job can be calculated as follows:

$$\frac{25{,}000 \text{ SF of TPO Membrane}}{30 \text{ SF per man hour}} = 834 \text{ Total man hours}$$

$$\frac{350' \text{ of wall flashing}}{3 \text{ LF per man hour}} = 116 \text{ MH}$$

Total man hours − 834 + 116 or 950 man hours.

The total duration for the job can be calculated using the crew size as follows:

The crew consists of five members. Each crew member will spend eight hours per day on the job site. This means the crew will collectively spend 40 hours daily on the job site. Therefore, the total number of days for the project will be

$$\frac{950 \text{ total man-hours}}{40 \text{ total man hours each day}} = 23.75 \text{ day}$$

10.6 Calculating Total Labor Cost

Assuming the same scenario, 25,000 SF of TPO membrane and 3′ wall flashing for 350′ with a five-man crew. The average labor man-hour cost will be:

$40 (one foreman) + $30 (one mechanic) + 3x$25 (three helpers/laborers)/5 = $29/hour

The total labor cost will be:

950 total man hours @ @29/hour average crew rate = $27,500.

There are several software programs available that are able to assist in roof estimating. However, it is essential to know the quantity, labor rates and productivity rates to calculate total man hours and total labor costs.

10.7 Calculating Supervision

Supervision refers to a non-productive person responsible for overseeing the work.

Supervision could be included in overhead costs or calculated separately.

From the previous example, the project consists of 24 crew days.

$$\frac{24 \text{ crew days}}{5 \text{ days a week}} = \text{total of five weeks}$$

It is essential to remember that the superintendent can become involved one week prior to the start of the project and most likely another week after the project is completed. Therefore, the total time for supervision in this example would be about 7 weeks or 280 hours.

Assuming the superintendent is paid $65/hour (from Figure 11.6.1), the total cost for supervision will be $65/hour x 280 hours or $18,200.

10.8 Mobilization and Demobilization

Mobilization and demobilization refer to the time needed to move equipment or supplies from and to the warehouse. Depending on the project's location, size and complexity, this could take anywhere from 60 to 400 hours.

Unloading and hoisting refers to the time needed to unload trucks, put material in storage and hoist it to the roof. Generally, this depends on the number of truckloads of material the project will need, but on average, 24 to 36 hours should be allowed for unloading and hoisting.

Travel time should also be included, especially getting to the site from the warehouse and picking up supplies and materials during work hours.

To summarize, an estimate must include materials and their quantities, labor, equipment, other subcontractor bids, general conditions, bonds, permit fees, taxes and profit.

Some other documents need to be provided along with the project estimate as requested by the owner, architect, or consultant. This is commonly known as the qualifications package and includes:

- Acknowledgment of all addenda
- Bonds
- Copy of certificate of insurance
- Financial statements
- Bidder qualification statement
- Projects similar in size and scope
- Project-specific safety plan

- Evidence of experience modification rate for insurance

- Letter from manufacturer accepting proposed roof system

- Letter acknowledging that you will sign the contract as presented

- Business License

- Evidence of use or attempt to use diverse suppliers

- List of all subcontractors that will be used

- Bid form

- References from banks, bonding companies, other clients, etc.

- How the bid documents are to be completed, delivered, etc.

Estimating requires extensive work. It is important to review all the documents, especially specifications and drawings, in detail before developing the estimate. Understanding all of the project's requirements is critical to producing an accurate estimate.

10.9 Estimating Tools

There are different methods of estimating depending on the project's size and complexity. Estimates can be done by hand; they can be done using Excel or other similar programs, and roofing-specific templates can also be purchased online.

10.10 Getting Started with Estimating

Estimating is an individualized task that is difficult to delegate. The good news is that estimating can start anywhere between plan review, specification review and front-end document review. The bad news is that all of these documents need to be reviewed by a single individual to get a comprehensive idea of the scope of the work.

CHAPTER
11

Safety in the Roofing Industry

with content donated by
Robert Pringle
CHST, CSE, CECM, NYSCSC
Evans Roofing Company, Inc.

11.1 Introduction

Roofing has always been considered dangerous work, mostly for the obvious reason it involves working at heights. According to the Bureau of Labor Statistics, there were 123 fatalities among roofing workers in 2021, and 99 of those resulted from falls – making it one of the most dangerous occupations in the country.

These statistics show how important it is to ensure safety in roofing projects. Safety involves training and educating employees, but it also requires continuous attention to the safety aspects of every job.

Comprehensive site inspections are critical to reducing accidents and injuries and keeping everyone on the job site safe. Standard tools needed for a safety inspection include:

- Tape measure
- Camera
- Safety inspection forms
- OSHA construction standards (hard copy, tablet or app)
- Electrical meter
- Personal protective equipment (PPE)

11.2 Identification of Hazards

There are two principal types of hazards that are common in the roofing industry: Existing hazards and predictable hazards.

1. Existing hazards are those that are currently present in the workplace and can potentially cause serious injury or death.

2. Predictable hazards are those that can appear in the workplace and potentially cause serious injury or death.

Predictable hazards can also be existing hazards. An example would be an older roof system that may contain asbestos, a known human carcinogen. The presence of asbestos can only be confirmed by having a core sample of the roof taken by an accredited testing agency. Once confirmed, necessary steps can be taken to mitigate the hazard.

The following figures indicate a few existing hazards to keep in mind. Figure 11.2.1 shows a roof with exposed roof edges. In this case, a fall protection plan should be developed, specifying what fall protection equipment or devices will be used. All crew members accessing the roof need to review and sign the fall protection plan.

Figure 11.2.1 – Roof with Exposed Edge
Source: OMG Roofing Products

Figure 11.2.2 shows a roof with a skylight that represents a fall hazard. In this case, the skylight must be inspected to see if the glass is fall-resistant. It if is not, a plan must be implemented to ensure that workers do not accidentally fall onto the skylight.

Figure 11.2.3 shows a roof with plastic corrugated panels installed to allow natural light into a building. Exposure to sunlight makes these panels brittle over time. This is an existing hazard that can lead to a fall if they are stepped on. Proper covers need to be installed to protect the workers.

Figure 11.2.4 shows an existing hazard for a hospital's laboratory, where exhaust fans for infectious diseases have been mounted on the roof. These fans need to be locked before work commences. In this case, it is essential to coordinate with the people managing the building's facilities since roofing work will be performed near the exhaust fans.

Figure 11.2.4 – Roof with Exhaust Pipes
Source: Evans Roofing Company, Inc.

Figures 11.2.5 show examples of hazards that are both predictable and existing. The roof, once torn open, can be seen to have a serious fall hazard. The site-specific safety plan needs to address this hazard.

Figure 11.2.5 – Existing and Predictable Hazards
Source: Evans Roofing Company, Inc.

11.3 General Requirements

A pre-job meeting should be held to ensure the involvement of all the stakeholders in the safety inspection. This initial meeting is vital to set the tone of safety culture and align everyone involved with the same goal of ensuring on-site safety. It is important to be sure general requirements, including the following, are met:

- Potable water

- First aid kit

- Fire extinguishers

- Safety Data Sheets and Job Hazards Analysis forms

- Emergency phone numbers

- Required posters

- Site-specific fall protection plan

- Toilet facilities

Some of these general requirements can be set up in a job site trailer or on the roof.

Several posters required by local, state and federal agencies must be on-site at a location

easily accessible to employees. An example is shown in Figure 11.3.1.

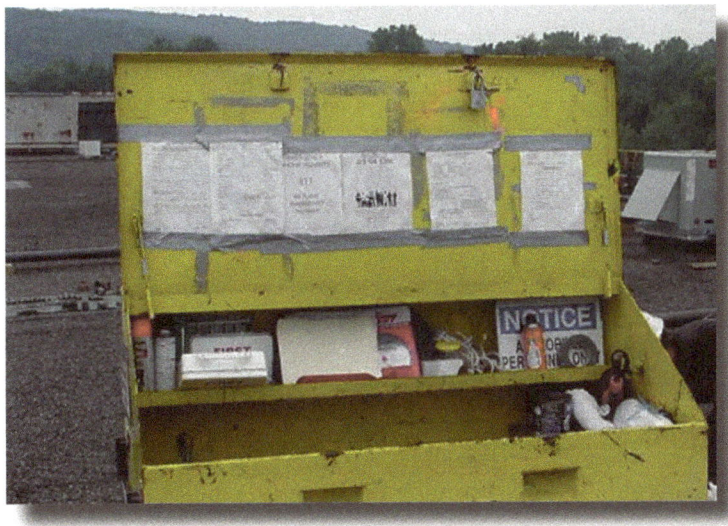

Figure 11.3.1 – Posters On Site
Source: Evans Roofing Company, Inc.

11.4 Site-Specific Fall Protection Plan

Figure 11.4.1 is an example of a Fall Protection Plan. In addition to fall protection options,

this form also includes emergency contact information and details about the nearest

medical facility.

Fall Protection Plan (FPP)

Project Name:		Revision #	
Address:		Prepared by	
City/State		Approved by	
Project #		COO/VP	
Date:		EH&S Dept.	

| Roof System: | | Roof Thickness: | |
| Slope: | | Type of Deck: | |

Fall Protection Options
Check Appropriate System(s)

Conventional			Alternative	
Guardrail System			Restraint System	
Warning Line System			Fall Arrest System	
Controlled Access Zone				

Description of Fall Protection System

Description of PFAS Fall Rescue Plan

Description of Roof Top Evacuation Plan

Emergency Contact # & Nearest Medical Facility

Figure 11.4.1 – Fall Protection Plan
Source: Evans Roofing Company, Inc.

11.5 Roof Access

There are multiple options for accessing a roof, including external portable ladders, external fixed ladders, scaffold towers, internal stairs, external stairs, roof hatches and elevators.

Portable external ladders can be dangerous. It is important to follow safe ladder practices:

- An external portable ladder must extend at least 3' higher than the access level, as shown in Figure 11.5.1.
- The ladder needs to be properly sloped, ideally at a ratio of 1:4.

Figure 11.5.1 – External Portable Ladder
Source: Evans Roofing Company, Inc.

In some cases, an external fixed ladder is affixed to the building and can be used if the owner allows it. Inspecting these ladders and ensuring they are in good working condition before workers use them is critical. One of the most common issues with fixed ladders is loose bolts that secure the ladder to the wall.

A scaffold tower, as shown in Figure 11.5.2, is set up for roof access when a fixed or portable ladder is not feasible. Once erected, the tower becomes a part of the inspection process and requires regular inspection to be sure it is secure and safe to use.

Figure 11.5.2 – Scaffold Tower
Source: Evans Roofing Company, Inc.

Figure 11.5.3 shows a good example of what a job site looks like when it has a good site logistics and safety plan. There is a scaffold tower, a trash chute, a portable toilet and a dumpster. The entire area is barricaded to keep people not involved with the project away.

Figure 11.5.3– Scaffold Tower, Trash Chute, Porta John, and Dumpster
Source: Evans Roofing Company, Inc.

Roof hatches are a good selection for access as long as they are thoroughly inspected before use. Stairs are another option for roof access when available, as shown in Figure 11.5.4. It is essential to have uniform treads and risers, properly anchored handrails, etc., to ensure safe use.

Figure 11.5.4– Stairs
Source: Evans Roofing Company, Inc.

11.6 Personal Protective Equipment

Personal protective equipment (PPE) includes eye and face protection, head protection, hand protection, vest/clothing, respiratory protection, foot protection and fall protection. Some typical fall protection equipment includes a retractable lifeline, rope grab and lifeline, four-point harness, rip stitch lanyard, slow-releasing decelerating stitched lanyard and carabiner hardware.

11.7 Fall Protection Systems

Motion Stopping Systems

This type of system protects a person falling from injuries, provided they are connected securely. The types of motion-stopping systems are as follows:

• Guardrail system. The construction of this type of system includes wood, metal, rope, chain and cable. The top rail of the system needs to be at least 42" high and able to withstand 200 pounds of applied force. The top rail should not deflect more than 2" between any two upright supports. There are many such specifications in OSHA standards. Figure 11.7.1 shows a guardrail system that will protect a person from falling off the roof's edge. Figure 11.7.2 shows an existing guardrail in place, and Figure 11.7.3 shows a guardrail installed with a parapet wall clamping system.

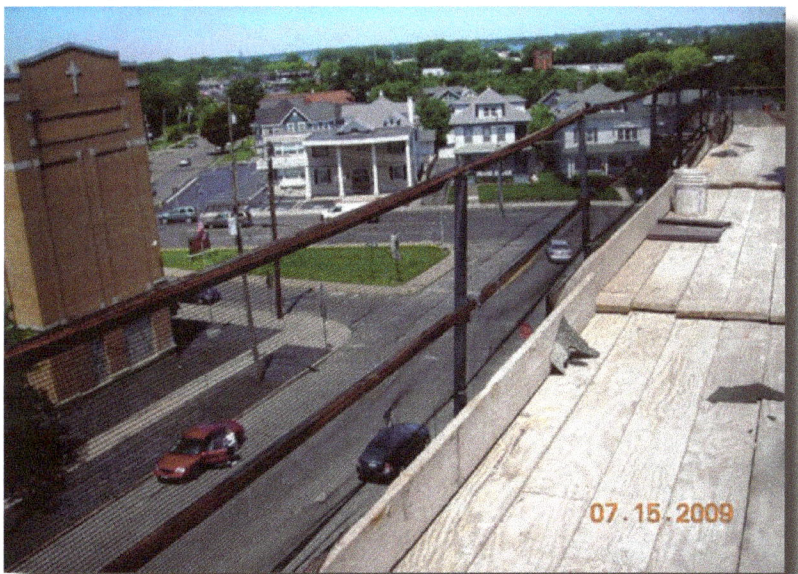

Figure 11.7.1– Guardrail System
Source: Evans Roofing Company, Inc.

Figure 11.7.2– Existing Guardrail in Place
Source: Evans Roofing Company, Inc.

Figure 11.7.3– Parapet Wall Clamp System
Source: Evans Roofing Company, Inc.

- Fall arrest system. The anchorage points and hardware of a fall arrest system must be capable of withstanding 5,000 pounds per employee. There should be one anchor point for each employee. All fall arrest components need to be inspected by a competent person. With this system, no employee must fall six feet or more, or be able to come in contact with any lower level or object below.

- Mobile fall carts. Employees need to be trained when they are used.

- Restraint system. The anchorage point for a restraint system must be able to withstand 3,000 pounds per employee. An inspection of all positioning system components for compliance needs to be conducted by a competent person. With this type of system, no employee must fall more than two feet or be able to come in contact with any lower level or object below. An example is shown in Figure 11.7.4.

Figure 11.7.4– Restraint System
Source: Evans Roofing Company, Inc.

- Covers. Any kind of opening on the roof needs to have a cover to prevent falls. OSHA requirements say that all covers must be capable of withstanding twice the intended load, including personnel, equipment and machinery. Covers need to be secured to prevent displacement and marked so workers can easily identify them. Figure 11.7.5 is a plywood cover that is secured to prevent movement. Figure 11.7.6 shows a non-typical cover over a skylight; it consists of framing and plywood. Figure 11.7.7 is a tabletop cover over a skylight that can be reused.

Figure 11.7.5– Plywood Cover
Source: Evans Roofing Company, Inc.

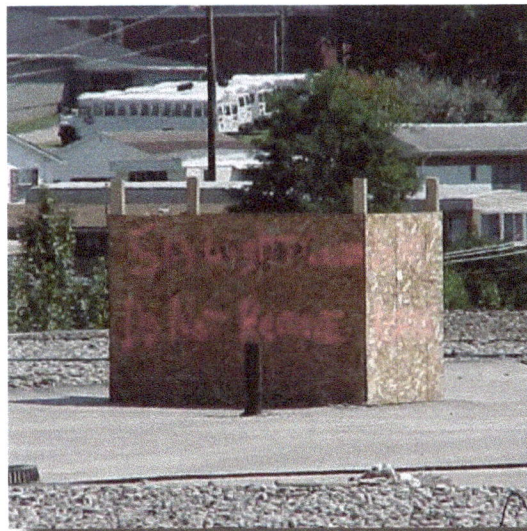

Figure 11.7.6– Non-topical Cover
Source: Evans Roofing Company, Inc.

- Parapet walls. OSHA requires that parapet walls be at least 39" above the roof surface and be structurally safe. A combination of parapet walls and guardrails can also be used for compliance. An example of a well-constructed parapet wall is shown in Figure 11.7.8.

11.8 Warning Line and Safety Monitor

OSHA regulations allow for the use of a warning line system and safety monitor on most low-slope roofs. The warning lines should be between 34" and 39" above the surface of the roof so they are clearly visible to workers.

The warning lines must be installed no less than 6' from the edge of the roof when operating equipment parallel to the edge and at least 10' from the roof edge when operating equipment perpendicular to the edge. The lines must be flagged a maximum of 6" with highly visible material. An example is shown in Figure 11.8.1.

Figure 11.8.1– Warning Lines
Source: Evans Roofing Company, Inc.

A Safety Monitor watches workers on the roof when they are inside the warning line and close to the edge of the roof. They must be competent in recognizing fall hazards. An example of a Safety Monitor at work is shown in Figure 11.8.2.

Figure 11.8.2– Safety Monitor
Source: Evans Roofing Company, Inc.

11.9 Hand and Power Tools

Hand tools are non-powered tools such as hammers, chisels, screwdrivers, and wrenches, as shown in Figure 11.9.1. Power tools include electrical, pneumatic, fuel, hydraulic, and powder. Electrical power tools are commonly used for roofing, as shown in Figure 11.9.2. Tools must be regularly inspected for compliance, and any defective tools must be replaced.

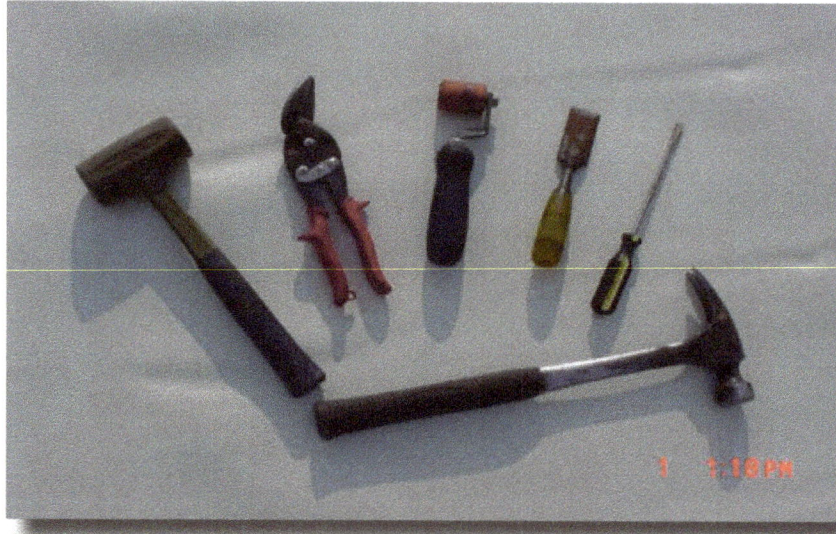

Figure 11.9.1– Hand tools
Source: Evans Roofing Company, Inc.

Figure 11.9.2– Power tools
Source: Evans Roofing Company, Inc.

11.10 Power Supply and Electrical Hazards

It is important to check for electrical hazards on the job site. Inspecting tools prior to use is critical to ensure they are in good working condition. Any tools found to be damaged or defective should be tagged for repair or replacement and not used.

There are three types of electrical power supply on a roofing project: the owner's direct power, utility pole power and generators. Regardless of the type, the power supply that is used must be inspected prior to use and must have a ground fault circuit interrupter (GFCI) installed between the power source and the tool. Any electrical equipment needs to be inspected before power starts running through it. Figure 11.10.1 shows a utility-powered panel box on the roof with GFCI breakers installed. When using portable generators, grounding cables are required to protect employees from electrical hazards. If there is an issue with the power source and/or generator, the grounding cable with direct the energy to the least path of resistance and away from the tool operator. Figure 11.10.2 shows a mobile generator-powered panel connected to a grounding cable on the roof.

Figure 11.10.1– Power Panel
Source: Evans Roofing Company, Inc.

Figure 11.10.2– Mobile Power Panel
Source: Evans Roofing Company, Inc.

OSHA requirements state that energized power lines must be a minimum of 10' from walking/working areas for 50KV and additional distances for greater voltage. The power line owner needs to be notified before the start of work. Power lines need to be insulated or blanketed for additional protection. A one-call system is followed to alert the utility company to the commencement of work so that they can send a team to apply the insulation.

11.11 Kettle Safety

For built-up and modified bitumen roof systems, it is crucial to ensure kettle safety. A roof kettle is used to heat asphalt for the application of these systems, and it should be properly barricaded. Liquid petroleum (LP) gas cylinders must be adequately secured and must be at least 10' away from the kettle. Storage of LP gas cylinders must be at least 20' from building doors and walkways. A backflow preventer and regulator must be installed. In Figure 11.11.1, the gas cylinder is placed 20' from the kettle with a fire blanket underneath it on the ground. The kettle is kept closed during operation.

Figure 12.11.1– Kettle Safety
Source: Evans Roofing Company, Inc.

The kettle operator must wear proper PPE with a full-face shield, cuffless cotton pants with leather lace-up boots and cotton gloves with tight-knit cuffs. If long-sleeved cotton shirts are worn, they need to have buttoned cuffs.

Ensure that the kettle's operating temperature is below the "flash point" for the bitumen that is being heated. Check with the bitumen suppliers for safe operating temperatures for their products. Additional precautions to ensure kettle safety are:

- Use lids that fit properly
- Use gauges that are in good working condition.
- Inspect hoses, clamps, gauges, regulators and fuel cylinders before use
- Have an adequate supply of water in the immediate area while heated bitumen is being applied
- Have an adequate number of fire extinguishers on-site
- Make sure the pipes, or "hot lines," reach the roof safely and do not obstruct the access path to the roof
- Make sure to eliminate any hazards in the barricaded area
- Make sure workers are trained about proper kettle use

Flammable storage areas should have signs indicating that flammable material is present. There should be fire extinguishers in the immediate area, and the area should be separated from other parts of the job site. They should be at least 10' from any means of egress and exit. Figure 11.11.2 is an example of a storage area for flammable materials.

Figure 11.11.2– Flammable Storage Area
Source: Evans Roofing Company, Inc.

11.12 Dumpster and Chute Areas

Dumpster areas need to be properly barricaded. Dumpsters should not get overloaded, and the area surrounding the dumpster should be kept neat and orderly. Signage should be used if required. Figure 11.12.1 shows the dumpster area for a large reroofing project.

Figure 11.12.1– Dumpster area
Source: Evans Roofing Company, Inc.

Trash chute areas need to be adequately secured. Fall protection needs to be installed and inspected daily near the chute's access point on the roof. The area should be barricaded to prevent any hazards created due to debris. Figure 11.12.2 is an example of a trash chute.

Figure 11.12.2–Trash Chute
Source: Evans Roofing Company, Inc.

11.13 Material Storage and Handling

Materials should be stored and barricaded properly. They should be covered to protect against the weather. If materials are stored on the roof, they need to be placed within warning lines. Ensure that all trailers are secure. Every roofing material manufacturer specifies storage requirements for their materials; these must be followed. Figure 11.13.1 shows the storage of roof insulation bundles on the roof. Notice that the bundles are placed within warning lines, covered and neatly organized.

Figure 11.13.1– Material Storage on the Roof
Source: Evans Roofing Company, Inc.

Standard options for material handling include hoists, powered industrial cranes, ATVs, four-wheel carts and dollies.

Hoists need to be inspected before operation. Hoists need to have counterweights secured from displacement and properly marked. Ensure that the area around the hoist is barricaded. Figure 11.13.2 is an example of a hoist. A hoist's hook needs a "keeper plate" to prevent the load from bouncing off of the hook as it is being lifed. An example of a hook with a keeper plate is shown in Figure 11.13.3

Figure 12.13.2– Hoist
Source: Evans Roofing Company, Inc.

Figure 12.13.3– Hoist
Source: Evans Roofing Company, Inc.

In cases where a crane needs to be used for material handling, it is necessary to have a competent person who understands hand signals on the roof so the crane operator can be efficiently guided. There should be proper planning about where to load materials, and the area around the crane needs to be barricaded to protect people on the ground. Figure 11.13.4 is an example of a crane loading material to the roof. Notice that the person communicating with the crane operator is properly tied off on the roof, and another person is on the ground to ensure that materials are safely rigged and secured for transporting to the roof.

Figure 11.13.4 – Crane
Source: Evans Roofing Company, Inc.

Power industrial trucks (PITs) are also known as forklifts are commonly used for moving materials horizontally on a jobsite. The operator of a PIT must carry current proof of training. The operator must inspect the truck daily and carry proof of inspection. The operator must wear proper PPE. It is essential to train operators for each type and model of PIT, as they are all different.

CHAPTER
12

Roof Repair and Maintenance

with content donated by
Chris Huettig
KARNAK Corporation

12.1 Introduction

Good roofing practices extend beyond the installation of the roof. A roof must be continuously maintained to ensure it functions to – and even beyond – its expected lifespan. Repair and maintenance are critical aspects of roofing. Some common examples of roof repair are shown in the following figures.

Figure 12.1.1 shows a liquid-repaired roof where a liquid or sealant is used to repair and protect the roof. Figure 12.1.2 shows a heat-welded thermoplastic membrane repair, and Figure 12.1.3 shows a thermoset membrane repair.

Figure 12.1.2– Heat-Welded Thermoplastic
Source: KARNAK Corporation

Figure 12.1.1– Liquid-Repaired roof
Source: KARNAK Corporation

Figure 12.1.3– Thermoset Membrane Repair
Source: KARNAK Corporation

12.2 Reasons for Roof Leaks and Failures

Roofs can fail for any number of reasons. Some common reasons for failure are:

- Constant exposure to the elements
- Expansion and contraction of dissimilar materials over time. A roof system can include multiple materials – e.g., asphalt, plastic, rubber, metal, masonry, stone and wood --- all of which have different rates of expansion and contraction. Over time, the bonds between these materials weaken, causing leaks.
- Reduced resistance to wind uplift, which can result in fasteners loosening and materials dislodging
- Tenant improvements and improper repair by inexperienced professionals
- Poor installation or poor material quality
- Lack of proper drainage, resulting in ponding and clogged drains
- Weather events or other unexpected catastrophes

If not properly addressed, all of these problems could ultimately lead to total system failure.

12.3 Importance of Roof Repair and Maintenance

A roof is a critical part of every building, and is often referred to as the "fifth wall." Roof leaks not only can indicate damage to the roof system but they can also:

- Damage insulation and reduce its efficacy
- Cause structural damage
- Rust steel supporting members
- Create mold and lead to unhealthy conditions for the occupants of the building
- Damage interior contents or valuable inventory
- Lead to premature roof replacement

While roof repair and maintenance can help prevent problems, keeping a roof properly maintained can also:

- Reduce a building's peak cooling demand by 10-15% by keeping insulation dry and maintaining good solar reflectivity. In some markets, this can lead to utility rebates and tax incentives.

- Maintain the fire rating of the roof, helping to keep insurance costs lower. Proper maintenance can also provide or maintain a National Sanitation Foundation (NSF) rating, which indicates whether the water that runs off the roof's surface is safe for use.

- Avoid the need for premature roof replacement.

12.4 Conducting Visual Inspections

The visual inspection of a roof system should begin with the interior of the building, to see if there is any mold, mildew, drips, puddles, water stains, peeling paint, stained ceiling tiles or other indications of a roof leak. If possible, the underside of the roof deck should be looked at for any signs of leaks or erosion. An example of a drop ceiling with water stains, indicating a potential roof leak, is shown in Figure 12.4.1.

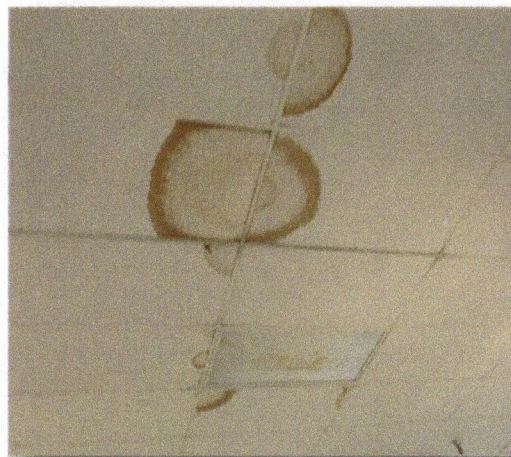

**Figure 12.4.1– Drop Ceiling with Water Stains
Indicating Potential Roof Leak**
Source: KARNAK Corporation

Next, it is important to walk the perimeter of the building on the ground and inspect the building's walls. Keep an eye out for any cracks in the façade, missing bricks or any damage to the exterior walls, as this can create additional stress on the roof system and lead to failure.

Before conducting a visual inspection of the roof itself, it is critical to have a plan and the proper safety equipment. Safety should always be the top priority. Before the inspection, understand how to access the roof. The building owner needs to be alerted before the roof is accessed.

When on the roof, an inspector must be aware of the surroundings. Hazards such as skylights mounted flush with the roof or skylights without safety guards need to be properly identified. Personal protective equipment, such as boots, gloves, hard hats and safety harnesses, are a must. An inspector must understand ladder and roof hatch safety, and be mindful of rooftop equipment.

Some standard tools needed for a visual inspection are:
- Notebook or tablet to take notes
- Camera to document defects or issues found
- Grease or wax pencil to mark the area of any defect found (Figure 12.4.2)
- Probe with a rounded, pointed end (Figure 12.4.3)
- Measuring tape or ruler
- A pair of gloves

All of these tools should be placed in a backpack to keep hands free while accessing the roof.

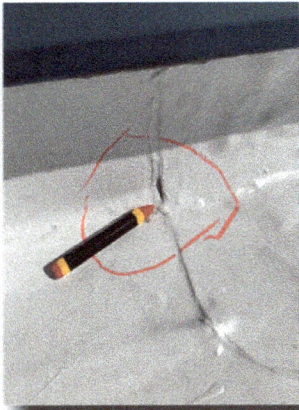

Figure 12.4.2– Grease Pencil Marking
Source: KARNAK Corporation

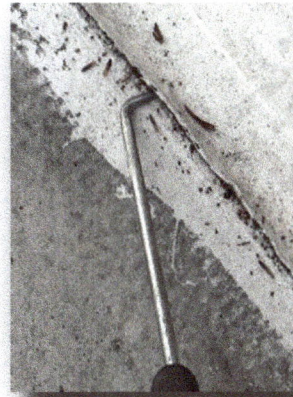

Figure 12.4.3– Probe
Source: KARNAK Corporation

Carrying a roof diagram, aerial imaging or Google Earth picture will help document issues observed on the roof and their locations. An example is shown in Figure 12.4.4. This will help in report writing once the inspection is completed.

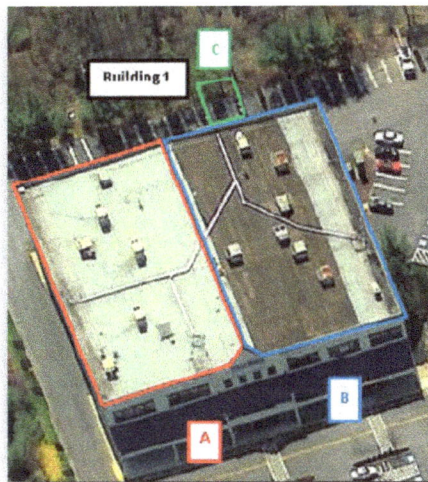

Figure 12.4.4– Roof Diagram
Source: KARNAK Corporation

It is also important to document the building's contact name, title, phone number and email address. When conducting a visual inspection, it is advisable to photograph the front of the building with signage to avoid confusion with other buildings – or roofs – that may look similar. If a photo of the roof is not available, a roof diagram can be made using graph paper, as shown in Figure 12.4.5. Make sure to include a general inspection checklist consisting of different roof sections and details in order to make notes during the inspection (see Figure 12.4.6).

Roof Plan Grid

No. _____

For use in conjunction with the Building Owner Maintenance Inspection Checklist

Building: _____ Inspection Date: _____

Location: _____ Inspector: _____

_____ Scale: 1" = _____

Figure 12.4.5 – Roof Plan Grid

Source: Manual for Inspection and Maintenance of Steep-slope Architectural Metal Panel Roof Systems; A Guide for Building Owners, NRCA, Rosemont, IL, 2003.

Inspection Checklist

Perform an initial general building inspection; conditions may indicate a roof system problem. Note the location for investigation on the roof.

ITEM		CONDITION SEVERITY: G = Good, No Action / F = Fair, Monitor Periodically / P = Poor, Immediate Action			
		G	F	P	ACTION TAKEN OR RECOMMENDED
ROOF SYSTEM CONDITION					
GENERAL	Debris				
	Walkways				
	Substrate Purlins				
	Contaminants				
	Leaks				
DRAINAGE	Roof Drains				
	Scuppers				
	Gutters				
	Downspouts				
	Ponding				
METAL ROOF PANELS	Seams/Joints				
	Loose Panels				
	Worn Panels				
	Damaged Panels				
	Fasteners and Washers				
	Fastener Holes				
METAL WALL PANELS	Seams/Joints				
	Loose Panels				
	Worn Panels				
	Damaged Panels				
	Fasteners and Washers				
	Fastener Holes				
FINISH	Worn Spots				
	Exposed or Corroded Metal				
	Adhesion				
	Cracks				
	Pinholes				
FLASHINGS	Roof-to-wall Flashings				
	Base Flashings				
	Counterflashings				
	Coping				
	Ridge Caps				
	Hips Caps				
	Valleys				
	Expansion Joints				
PENETRATIONS	Pipes				
	A/C Units				
	Vents				
	Skylights				
	Access Hatch				
	Ducts				
OTHER					

General Remarks: _____

Figure 12.4.6 – Inspection Checklist

Source: Manual for Inspection and Maintenance of Steep-slope Architectural Metal Panel Roof Systems; A Guide for Building Owners, NRCA, Rosemont, IL, 2003.

There are a number of technology options to assist in visual inspections. Inspection software generally improves reporting and saves time, and there are several applications available for mobile devices. The use of drones helps provide a quick overview of the roof from the safety of the ground; however, manual visual inspections allow for a more detailed and close-up observation of the roof.

12.5 Where Roof Problems Typically Occur

Areas where leaks or other issues with the roof commonly occur at the perimeter of the roof, as well as near flashings, penetrations, drains, gutters, scuppers, seams, expansion joints, caulking and sealants and where there is ponding water. Pay special attention to all of these areas during a visual inspection.

Figure 12.5.1 is an image of a roof termination area with caulking. When caulk is used with terminating structures such as counterflashing, reglets and termination bars, those areas must be inspected and re-caulked as required.

Figure 12.5.1– Roof Termination at Parapet Wall
Source: KARNAK Corporation

Figure 12.5.2 is an example of a flashing area where roof deficiencies commonly occur. Multiple materials come together in such areas; they all move independently, and leaks can result.

Figure 12.5.2– Flashing
Source: KARNAK Corporation

Figure 12.5.3 shows a typical roof penetration. Since penetrations such as pipe vents are not attached to the roof, they are subject to vibration or movement, which can break down their seals. Most penetration details include sealants and/or caulk that need to be maintained and reapplied regularly.

Figure 12.5.3– Penetration
Source: KARNAK Corporation

Drains get clogged frequently and need to be inspected. An example is shown in Figure 12.5.4. Notice that there are no drain baskets, which often get displaced due to wind or water. In such cases, large pieces of debris could make its way down the drain, clogging it and leading to ponding on the roof.

Figure 12.5.4–Drains
Source: KARNAK Corporation

As shown in Figure 12.5.5, gutters also need special attention. Clogged gutters can result in water accumulating on the roof, and additional damage can occur in the winter if the water freezes and thaws.

Figure 12.5.5– Gutters
Source: KARNAK Corporation

Scuppers, as shown in Figure 12.5.6, like drains, need to be inspected for blockages and cleaned out as needed to allow water to properly flow off the roof.

Figure 12.5.6– Scuppers
Source: KARNAK Corporation

Seams in the membrane or at flashings can fail over time for any number of reasons: exposure to weather, UV radiation, freezing and thawing, foot traffic and heavy traffic, just to name a few. Seams need to be checked regularly to ensure they are watertight. An example of a failed seam is shown in Figure 12.5.7.

Figure 12.5.7 – Seams
Source: KARNAK

As shown if Figure 12.5.8, expansion joints – which join two different roof sections and consist of elastic materials -- are subject to fatigue stress and need to be checked regularly. Over time, these materials lose their elasticity and can result in roof leaks.

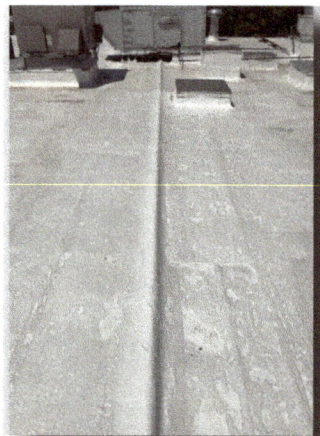

Figure 12.5.8 – Expansion Joints
Source: KARNAK Corporation

Caulking and sealants are transitional materials often used between different types of materials. Stress from the substrate moving independently and exposure to the elements can cause caulking and sealants to wear down. They need to be maintained and reapplied throughout the life of the roof system. Figure 12.5.9 shows caulking at a counterflashing and coping stone. Figure 12.5.10 shows caulking along counterflashing and between a parapet wall and a brick wall.

Figure 12.5.9 – Caulking at Counterflashing
Source: KARNAK Corporation

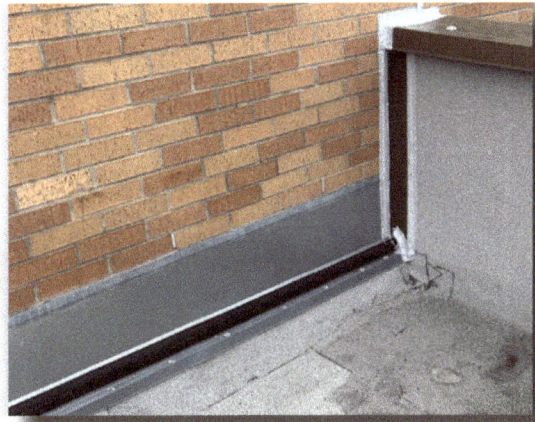

Figure 12.5.10 – Caulking at Parapet Wall
Source: KARNAK Corporation

Ponding water on a roof can cause a number of problems, such as dirt accumulation and algae growth, as shown in Figure 12.5.11. In addition, ponding water can also add significant weight to the roof, which may result in the deflection of a metal deck. A 5' by 5' ponded area with a depth of 2" can weigh more than 250 pounds.

Figure 12.5.11 – Ponding
Source: KARNAK Corporation

Building codes require that roofs be designed with a minimum of 2% slope, which equates to ¼:12. However, coal tar pitch roofs are an exception to this rule. Though rarely used any longer, coal tar pitch is not affected by ponding water the way other roof systems are. NRCA says the criterion for judging the proper slope of a roof is that there should be no ponding water 48 hours after rain and during conditions conducive to drying.

HVAC or other equipment placed on the field of the roof, as shown in Figure 12.5.12, could have its panels dislodged during extreme weather conditions. The sharp edges of the panels could wind up damaging the roof.

Figure 12.5.12– Loose Equipment on the Roof
Source: KARNAK Corporation

It is common to see fasteners backing out of the roof system, as shown in Figure 12.5.13. These fasteners can puncture the membrane and result in water penetrating the roof system.

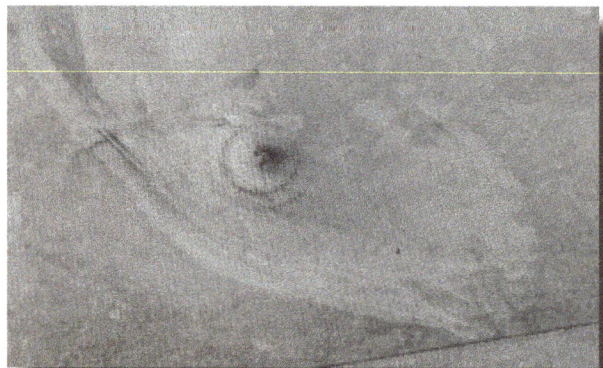

Figure 12.5.13– Fastener Backing Out
Source: KARNAK Corporation

Screws and nails that fall on the roof, as shown in Figure 12.5.14, could accidentally be stepped on, resulting in a puncture and a source of water entry.

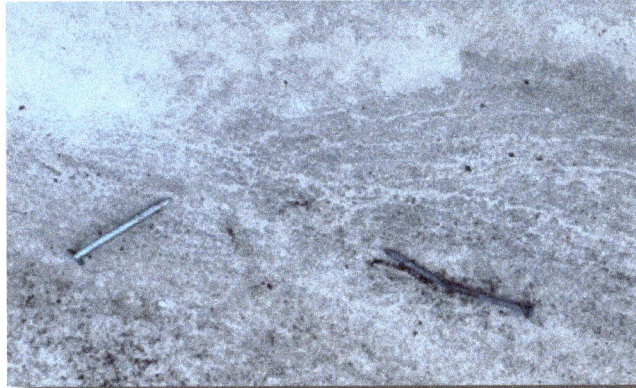

Figure 12.5.14– Screws and Nails
Source: KARNAK Corporation

Years of exposure to sun, wind, rain and snow can cause the roof membrane itself to deteriorate, as shown in Figure 12.5.15. While initially, this damage may only be on the roof's surface, prolonged exposure will cause these cracks to deepen and let water penetrate the system.

Figure 12.5.15– Deteriorating Membrane
Source: KARNAK Corporation

12.6 Finding Roof Leaks

Evidence of water inside a building rarely translates to a leak directly above the roof. Once water enters the roof system, it flows down the slope until it finds a point of entry into the building, such as a penetration. The source of the leak is often located many feet away from the point of water entry. Roof systems also absorb moisture until saturated, making it even more challenging to find the source of the leak and the extent of the damage.

There are three non-destructive methods for finding roof leaks:

- Nuclear gauges which are neutron-generating tools that move in a grid pattern across a roof's surface. They emit high-energy, fast-moving neutrons into the target area. These neutrons interact with the atoms in the roofing material and are slowed down by the presence of hydrogen (one component of water). The gauge calculates the number of slow-returning neutrons to find the moisture content of the roof.

- Capacitance induction equipment, which, like nuclear gauges, is moved in a grid pattern on the roof's surface. It uses the principle of induction to create a low-frequency alternating current between two electrodes, and determines the amplitude of the current detected, enabling it to calculate moisture content.

- Infrared thermal imaging, using an infrared camera to detect heat. The camera is usually used at dusk because the roof absorbs solar heat during the day and emits heat at the end of the day. Moisture within the roof system retains the heat for a longer period of time, so the infrared camera is able to detect the areas of the roof that are warmer and more likely to contain moisture.

All of these non-destructive tools are used to assist in a visual inspection to locate problems on the roof. Figure 12.6.1 is an example of a grid layout for a moisture survey. The figures show an outline of the roof with coordinates to help identify its different areas. The scale for this test ranged from 1 to 100. Areas with a high moisture reading indicated in red in the figure, will range from 40 to 100. Areas with a moderate amount of moisture, shown in green, will range from 30 to 39. Areas indicated in yellow have low moisture and will have a reading between 20 and 29. Areas with readings less than 20 have no discernable moisture. These areas are still given numbers due to hydrogen present in the roof materials. The testing is then validated by taking a core sample of the areas in red and green.

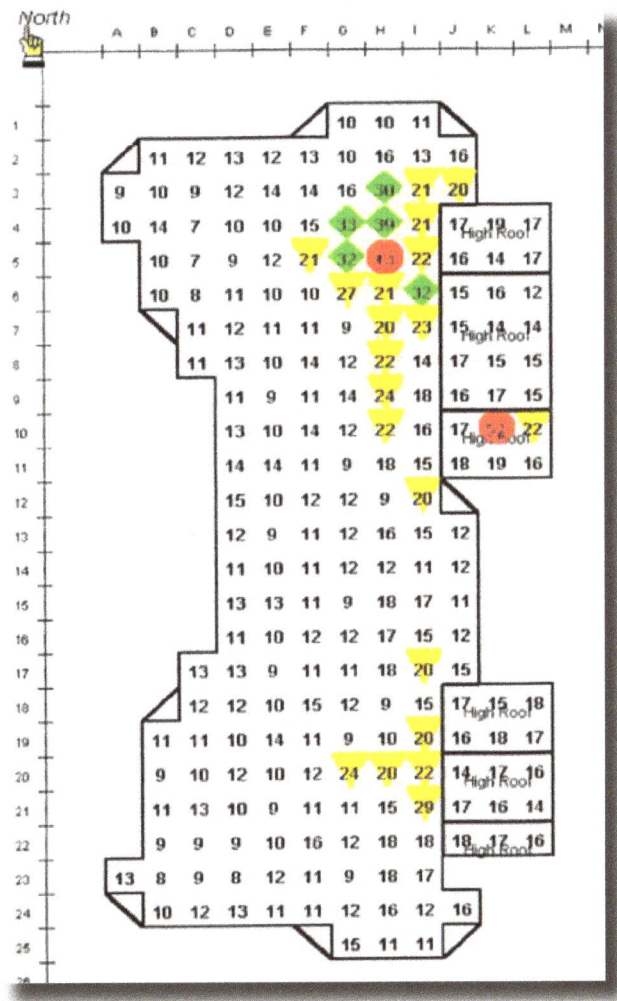

Figure 12.6.1– Gridded Roof Layout for a Moisture Survey
Source: KARNAK Corporation

Infrared imaging provides images, as shown in Figure 12.6.2. These might be black and white or color images. In the figure, black indicates the cooler areas, and white indicates the warmer areas, thereby showing the location of moisture.

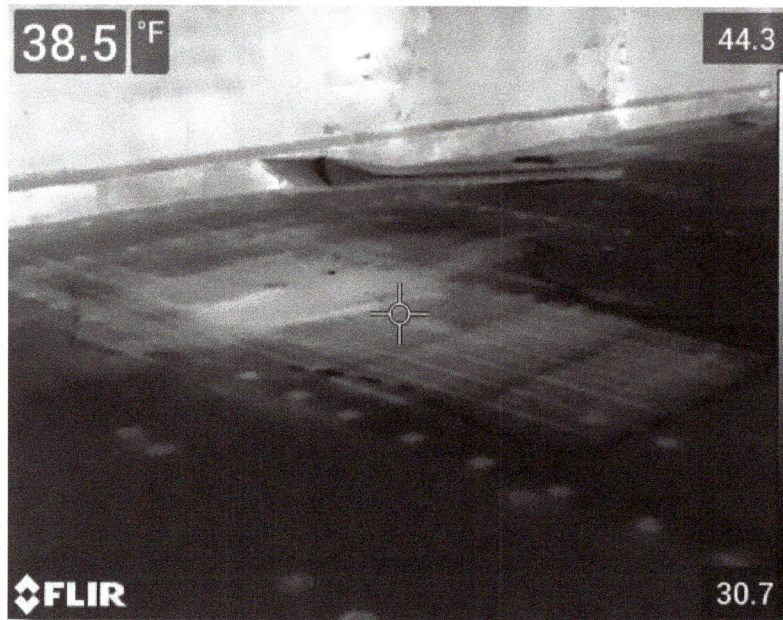

Figure 12.6.2– Infrared Imaging
Source: KARNAK Corporation

In addition to a visual inspection, it is usually important to conduct a non-destructive test. These tests are normally conducted twice, using two different methods to ensure accuracy. Following the testing, a core cut is taken to validate the results. These steps are all important to determine where leaks might be occurring and to evaluate the overall condition of the roof system.

12.7 Protective Roof Maintenance

Roof leaks that are not attended to can lead to a failure of the total roof system. Leaks usually arise from a lack of preventive roof maintenance. It is critical to have a routine maintenance plan with semi-annual inspections to identify leaks and any other defects that may be present – and then to perform any corrective action that may be needed.

The best approach to managing a roof system is to be proactive, and not just react to problems. In too many instances, if a building owner simply reacts to an issue once it has developed, damage may have been done to other areas of the roof. In addition, a roof warranty typically will not cover repair expenses if a lack of maintenance caused the defect.

Some of the advantages of having proactive roof maintenance are:

- Avoiding expenses outside of the normal budget cycle

- Extending the service life of the roof system

- Keeping the roof under warranty

- Preventing damage to equipment and inventory by finding and preventing leaks

- Ultimately reducing the total cost of the roof over its lifetime

Proactive roof maintenance typically costs about 1-3% of the estimated total roof replacement cost. Proactive maintenance typically includes most or all of the following:

- Cleaning drains, gutters and scuppers to allow for the free flow of water

- Cutting back overhanging trees

- Reapplying caulks and sealants

- Reconnecting condensate lines as needed

- Power washing to remove dirt, debris and contaminants

- Picking up nails, screws and other debris

- Performing minor repairs

- Documenting details of who has accessed the roof and what work has been done (this has to be done by the owner)

CHAPTER

13

**Using Capture Imaging
Technology in Roofing**

with content donated by

Piers Dormeyer

EagleView

1.1 Specialized Capture Imaging

Image capture technology is widely used in the roofing industry as it drives property reports and data. The data procured through this method adds great value to the roofing or reroofing process. Specialized capture imaging refers to the use of specialized techniques to capture images. Figure 13.1.1 shows an example of specialized capture imaging using equipment attached to a plane. This type of capturing device is widely used to collect images of high resolution (2 to 4 inches per pixel) compared to satellite images (18 inches per pixel), as shown in Figure 13.1.2. Aerial photography clearly captures the roof profile and roof membrane details.

Figure 13.1.1 – Specialized Capture Imaging
Source: EagleView

Figure 13.1.2 – Satellite Images vs. Aerial Images
Source: https://pxhere.com/en/photo/1552945 (CC0)

In addition to capturing images, this technology can be used to perform "QuickSquares" measurements (Figure 13.1.3), roof measurements, commercial measurements (Figure 13.1.4) and wall measurements.

Figure 13.1.3 – QuickSquares Measurement
Source: EagleView

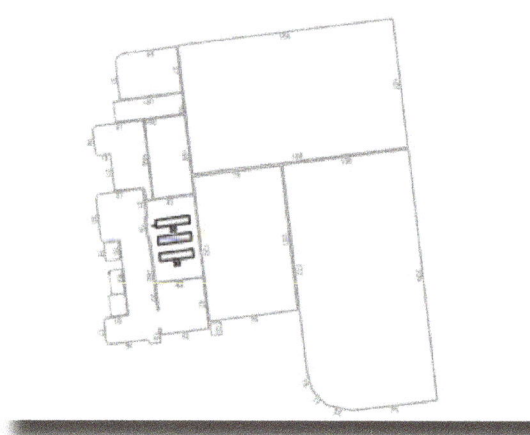

Figure 13.1.4 – Commercial Measurements
Source: EagleView

When compared with satellites and drones, the major advantages of using specialized capture imaging are:

- Image quality and resolution and

- Remote application (no site visit required)

The primary disadvantage is that geographic coverage is limited in rural or remote areas.

Resolution is measured by Ground Sample Distance (GSD), which refers to the distance between pixel centers measured on the ground. GSD determines the resolution of the imagery and the accuracy of the extracted data.

GSD imagery below 4" is ideal for accurate data extraction. A few examples are shown in the figures below. Figure 13.1.5 is a satellite image with a GSD =>15." Figure 13.1.6 is a specialized capture image with a GSD <3," and Figure 14.1.9 is a drone image with a GSD =<3." Both the specialized capture and the drone images have a higher resolution; however, drones require on-site presence.

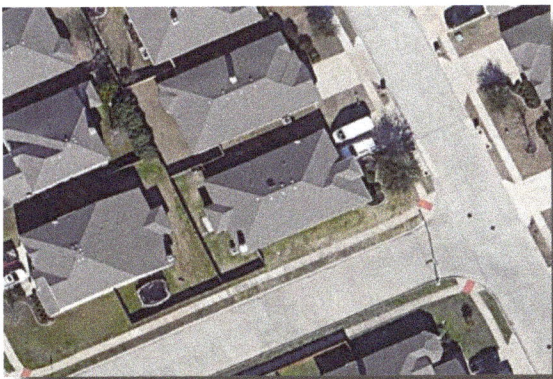

Figure 13.1.5 – Satellite Image
Source: EagleView

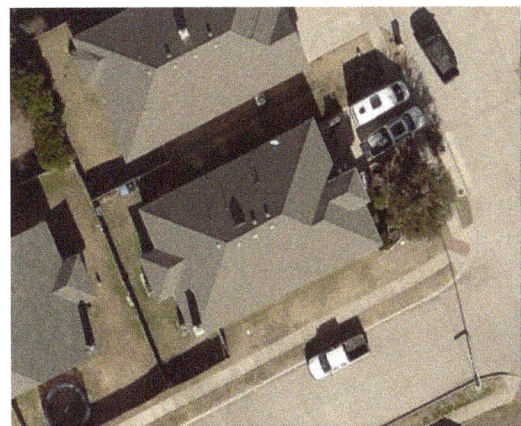

Figure 13.1.6 – Image Capture
Source: EagleView

Figure 13.1.7 – Drone Image
Source: EagleView

Orthogonal views provide a top-down view and are great for determining horizontal features on earth and structures. However, to get a better understanding, different perspectives are needed. Image capture technology allows the capture of oblique views that are useful to determine both horizontal and vertical features on earth and structures. Figure 13.1.8 is an example of an orthogonal view, and Figure 13.1.9 is an oblique view.

Figure 13.1.8 – Orthogonal view
Source: EagleView

Figure 13.1.9 – Oblique view
Source: EagleView

13.2 The Use of Technology

The advantages of using this technology include:

- Creating accurate bids quickly
- Professional estimates and proposals
- The opportunity for closing a sale with a single visit
- Selling upgrades easily
- Accurate material orders
- Greater profitability
- Compressed sales cycles (from 2 weeks to 2 days)

Not using this technology can result in the following:

- Inaccurate bids
- Material shortages or overages
- Additional trips to the supplier
- Unexpected disposal or storage costs
- Project delays and backlogs

13.3 Property Reports

Companies, such as EagleView, provide detailed remote property reports through high-resolution imagery. The company offers products that include roof measurements, 3D models and other key installation attributes. Another of its products includes solar access values that provide insight into the best areas for placing solar panels to attain maximum energy output. The Premium Roof Report also includes details on the roof's pitch, total facets, total line lengths for ridges, hips, valleys, rakes and eaves, as well as other data relevant to roofing contractors for creating bids and designs. The company also offers a web-based 3D visualizer that auto-generates final, install-ready photovoltaic designs based on the most trustworthy data available.

Figure 13.3.1 is an example of a report showing solar access value measurement (12,000 to 13,000 points per roof) compared to Figure 13.3.2, which shows measurements gathered by a hand-held measuring device (5-15 points per roof). This level of detail is obtained by the generation of a 3D roof model with 3D obstructions using leaf-off imagery (imagery captured when trees are bare) and then developing a Digital Surface Model (DSM) using leaf-on imagery (captured when trees have leaves).

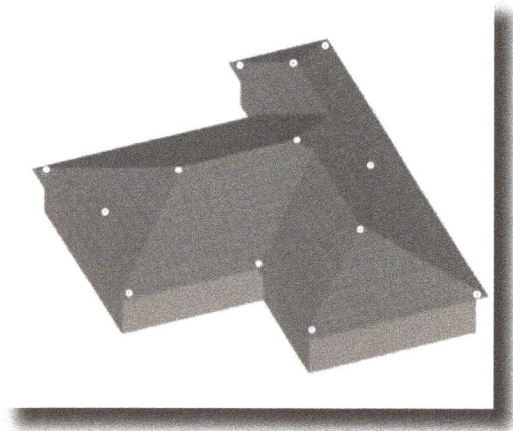

Figure 13.3.1– Fisheye Hand-held tool
Source: EagleView

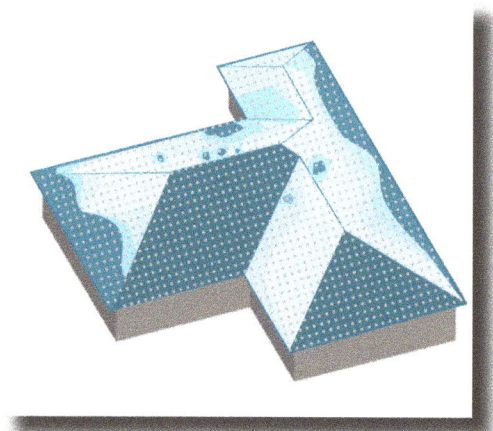

Figure 13.3.2 – EagleView Inform Advanced
Source: EagleView

After aligning relative and position, the 3D model is merged with the DSM. Following this, a six-inch grid is identified across the roof's surface, which creates thousands of measurement points. The path of the sun is mapped over this merged model. Solar access for each 15-minute interval throughout the year is recorded for each measurement point, leading to highly accurate data on solar access values for the roof.

The company also offers a drone-based measurement tool that delivers consistent, high-quality property inspections with simple-to-fly drone technology. These are autonomous flights that provide accurate measurements and anomaly detection. They produce a digital version of the entire roof with automatic detection for every point of damage, which will be useful to provide to insurance companies.

A typical project cycle for a residential roof takes 4-6 weeks to complete. Scheduling exterior visits with homeowners typically consumes a significant amount of time. Studies have shown that site visits take 2-3 hours and cost at least $250. With the use of image capture technology, this step can be eliminated to save time and increase productivity.

Another critical reason to implement this technology is to simplify the design process. Site measurement error is a common problem that slows the design process and leads to change orders. Sometimes, these change orders can also affect the schedule and cost significantly. Statistics show that 9 out of 10 companies do a redesign of their project at least once. Image capture technology can help eliminate the need for an exterior site visit by providing consistent, accurate roof measurements and details. It can help reduce the number of change orders and re-work, thereby contributing to cost and time savings. EagleView project cycle time claims to reduce project costs by up to 75% and completion times by more than 50%.

These reports are available not only in PDF formats but also DXXF and XML formats, making it easier and quicker to integrate the systems that roofing contractors and solar installers typically use.

13.4 Using Technology to Engage with Customers

Salespeople need to be well prepared before engaging with customers-to-be to be able to influence their decisions. There are apps that can help organize tasks beforehand, thereby reducing the number of visits and improving productivity.

The application helps with the following:

- Quick qualification of leads and assigning them to teams
- Tracking the progress of projects and the performance of the team

- Keeping tabs on scheduled appointments

- Validating property addresses after inputting homeowner details

- Collecting accurate, detailed data for estimates and installations without a site visit

- Maintaining a library of digital documents

The application's presentation mode, as shown in Figure 13.4.1, helps display credibility by showcasing designs and options. Proposals can include multiple material choices and 3D visualizations. Every bid can be customized with estimating tools that help showcase different quality and cost package options to the client with minimal effort. Accurate estimates with customizable templates and pre-populated pricing provide consistency and allow flexibility to make adjustments.

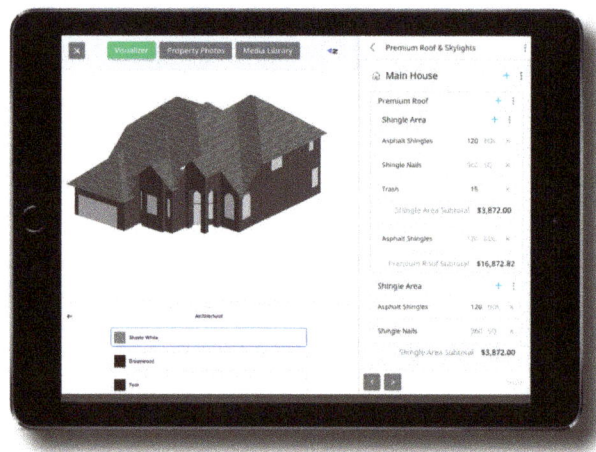

Figure 13.4.1– Example of EagleView Application Presentation Mode
Source: EagleView

The application also provides data analytics that help with managing leads and tracking proposals. It also produces realistic 3D models of projects delivered from high-resolution aerial oblique intelligent images and provides accurate 3D building representation. Figure 13.4.2 shows a 3D model of Greenville, South Carolina. Figure 13.4.3 shows a 3D model of the Atlanta airport.

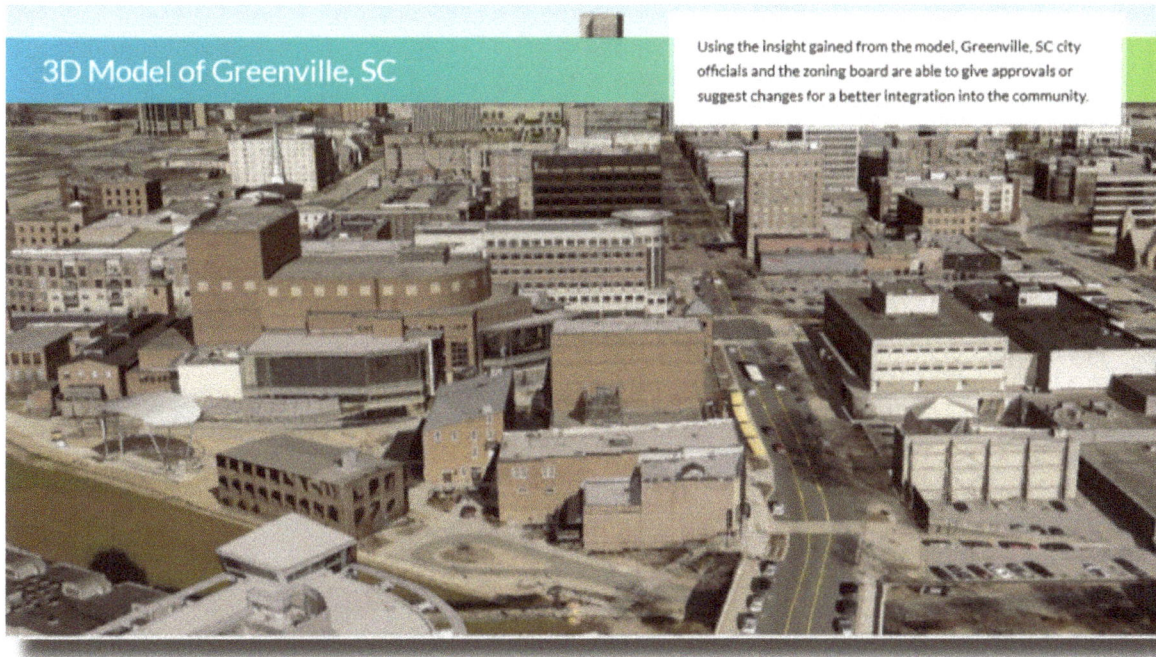

Using the insight gained from the model, Greenville, SC city officials and the zoning board are able to give approvals or suggest changes for a better integration into the community.

3D Model of Greenville, SC

Figure 13.4.2 – 3D Model of Greenville
Source: EagleView

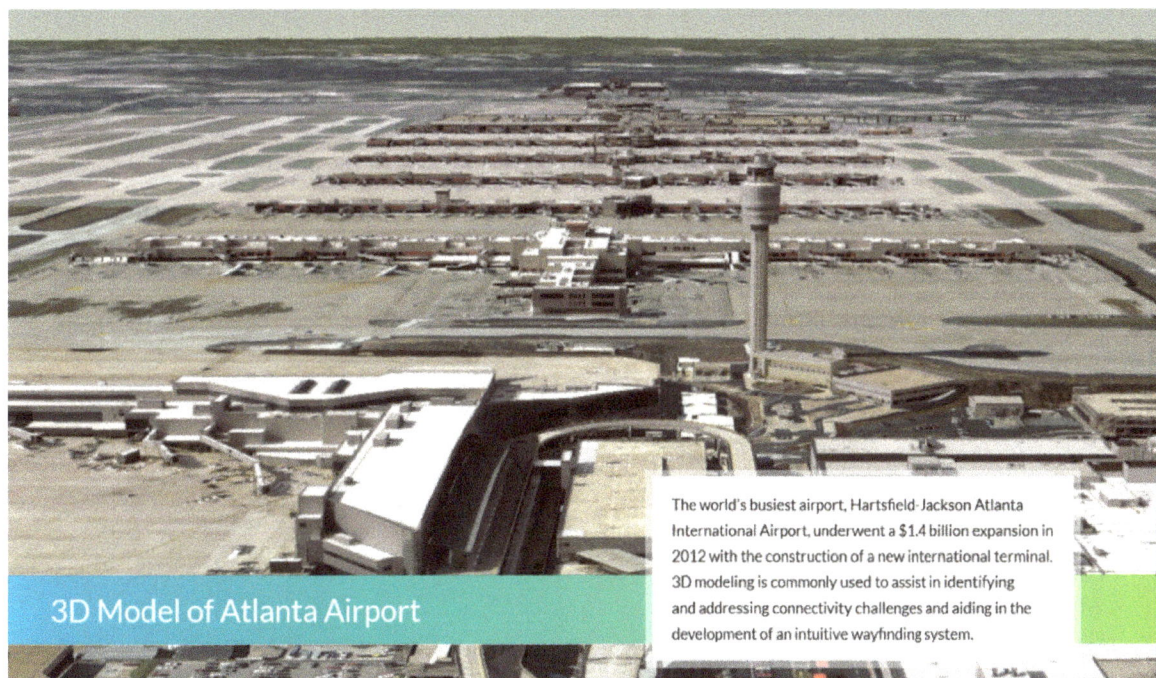

The world's busiest airport, Hartsfield-Jackson Atlanta International Airport, underwent a $1.4 billion expansion in 2012 with the construction of a new international terminal. 3D modeling is commonly used to assist in identifying and addressing connectivity challenges and aiding in the development of an intuitive wayfinding system.

3D Model of Atlanta Airport

Figure 13.4.3 – 3D Model of Atlanta airport
Source: EagleView

The path to success in the construction industry now relies on technology. Changing buying patterns among homeowners has driven innovation in the industry. Appointments are being converted to virtual ones that align with current buying ideologies.